Praise for
Supply Chain Network Design

"Due to the significant impact that supply chain design has on the cost and service provided by a company, it is critical that managers be knowledgeable about how to optimize the flow of products and goods within their supply chain. This book takes a very technical subject and makes it possible for managers and students alike to understand all aspects of network design. The practical approach used in discussing topics throughout the book provides a clear and excellent framework for those seeking to learn more about the topic. The book is a needed and welcome contribution to the profession."

—**Dr. Mary C. Holcomb**, Associate Professor of Logistics, Department of Marketing and Supply Chain Management, College of Business Administration, University of Tennessee

"Supply chain management (SCM) is a rapidly growing area of study—and network design is one of the fastest growing areas within SCM. This book would make a great classroom textbook. It is easy to follow with clear examples and useful exercises. It is laid out in progressive layers, with one's understanding of the subject matter building as you go. You can stop midway and be comfortable with the fundamentals, or you can go as deep as desired into *Supply Chain Network Design*.

"I have been a long-time practitioner of network modeling, as a manager in business and as a consultant, and I have covered the topic in university lectures; I still learned a great deal about a subject I thought I knew thoroughly!"

—**Bill Nickle**, Principal, Nickle Consulting

"This is a wonderful book that offers students and practitioners the opportunity to learn how to use quantitative techniques to design a supply chain from experts in the field. Along with covering how to model various issues in supply chain design (multiple echelons, service levels, and so on), the authors draw on their extensive expertise as practitioners to provide valuable insights into how to successfully execute a network design study."

—**Dr. Mike Hewitt**, Assistant Professor, Industrial and Systems Engineering, Kate Gleason College of Engineering, Rochester Institute of Technology

"*Supply Chain Network Design* will help students, academics, and practitioners alike understand the importance of successfully designing and optimizing a global supply chain network, while also explaining in easy to understand steps how to make it happen."

—**John A. Caltagirone**, Lecturer and Executive Instructor of Global Supply Chain Management, Quinlan School of Business, Loyola University

"This is the first book I have seen that starts with the basics of facility location, and then takes a step-by-step approach to adding real-life complexities to the modeling effort. It appropriately emphasizes the complexities and nuances of designing a supply chain network without getting bogged down in too much math."

—**Dr. Keely L. Croxton**, Associate Professor of Logistics, Department of Marketing and Logistics, Fisher College of Business, Ohio State University

"The book is well-positioned to answer many of the 'what happens in the real world' questions my students often ask. The material bridges the gap between classroom models and data or implementation realities. The exercises provided in the book challenge students to analyze their modeling assumptions and consider the implications of these assumptions. The case studies provide a great level of practical relevance for students. This book is a valuable guide for both engineers and supply chain practitioners."

—**Karen Smilowitz**, Associate Professor, Department of Industrial Engineering and Management Sciences, Northwestern University

SUPPLY CHAIN
NETWORK DESIGN

SUPPLY CHAIN NETWORK DESIGN

APPLYING OPTIMIZATION AND ANALYTICS TO THE GLOBAL SUPPLY CHAIN

Michael Watson, Sara Lewis, Peter Cacioppi, and Jay Jayaraman

Vice President, Publisher: Tim Moore
Associate Publisher and Director of Marketing: Amy Neidlinger
Executive Editor: Jeanne Glasser Levine
Editorial Assistant: Pamela Bolad
Development Editor: Russ Hall
Operations Specialist: Jodi Kemper
Marketing Manager: Megan Graue
Cover Designer: Chuti Prasertsith
Managing Editor: Kristy Hart
Project Editor: Jovana San Nicolas-Shirley
Copy Editor: Cheri Clark
Proofreader: Debbie Williams
Indexer: Lisa Stumpf
Senior Compositor: Gloria Schurick
Manufacturing Buyer: Dan Uhrig

©2013 by Michael Watson, Sara Lewis, Peter Cacioppi, and Jay Jayaraman
Publishing as FT Press
Upper Saddle River, New Jersey 07458

FT Press offers excellent discounts on this book when ordered in quantity for bulk purchases or special sales. For more information, please contact U.S. Corporate and Government Sales, 1-800-382-3419, corpsales@pearsontechgroup.com. For sales outside the U.S., please contact International Sales at international@pearsoned.com.

Company and product names mentioned herein are the trademarks or registered trademarks of their respective owners.

Printed in the United States of America

First Printing: August 2012

ISBN-10: 0-13-301737-0
ISBN-13: 978-0-13-301737-3

Pearson Education LTD.
Pearson Education Australia PTY, Limited.
Pearson Education Singapore, Pte. Ltd.
Pearson Education Asia, Ltd.
Pearson Education Canada, Ltd.
Pearson Educación de Mexico, S.A. de C.V.
Pearson Education—Japan
Pearson Education Malaysia, Pte. Ltd.

The Library of Congress cataloging-in-publication data is on file.

To my wife and kids for their support
while I was writing this book.
MSW

To my parents and siblings for their encouragement
and pride in all that I do.
And to my coauthors of this book, all the lessons,
debates, and work we have done together
will benefit me for many years to come.
SEL

To David and Edith Simchi-Levi,
who had the vision (and the aplomb) to allow me to work
with networks (and businesses) in the "real world."
PJC

To my family for their motivation and support
through this exciting journey.
JJ

CONTENTS

Part II: Adding Costs to Two-Echelon Supply Chains

6
ADDING OUTBOUND TRANSPORTATION TO THE MODEL 99

7
INTRODUCING FACILITY FIXED AND VARIABLE COSTS 127

8
BASELINES AND OPTIMAL BASELINES 139

Part III: Advanced Modeling and Expanding to Multiple Echelons

9
THREE-ECHELON SUPPLY CHAIN MODELING 157

10
ADDING MULTIPLE PRODUCTS AND MULTISITE PRODUCTION SOURCING 177

11
MULTI-OBJECTIVE OPTIMIZATION 207

Part IV: How to Get Industrial-Strength Results

Part V: Case Study Wrap Up

ACKNOWLEDGMENTS

This book was made possible through our work with the network design application LogicNet Plus XE, which was created by LogicTools and acquired by ILOG before it became part of IBM in 2009. The authors of this book have been associated with this product in some way since 1997.

Throughout this time, we have had the opportunity to work with an extraordinary team of colleagues and customers who have given us invaluable insight into the practice and theory of supply chain network design. This book would not have been possible without this experience.

While we cannot thank everyone we worked with, we would like to take the time to acknowledge some of our colleagues.

First and foremost, thanks to Derek Nelson for all his modeling and design expertise throughout the years. And a special thanks to David and Edith Simchi-Levi for founding and nurturing LogicTools, which ultimately made all of this possible. We learned a lot from both David and Edith.

Our extremely talented modeling team has not only helped to enable our customers, but also given us great insight into network modeling across all industries around the globe. The leaders of this team include Ganesh Ramakrishna, Remzi Ural, Alex Scott, Amit Talati, Guru Pundoor, Sumeet Mahajan, Nikhil Thaker, Ritesh Joshi, Javid Alimohideen, Shalabh Prasad, Ronan O'Donovan, and Dan Vanden Brink. Others on this team who have contributed in past years include Sharan Singanamala, Derrek Smith, Ali Sankur, Orkan Akcan, Louis Bourassa, Shaishav Dharia, Bhanu Gummala, Kristin Gallagher, Aziz Ihsanoglu, Giray Ocalgiray, Marcus Olsen, Ram Palaniappan, Chuin Kuo, Sanjay Sri Balusu, and Steven Lyons.

Besides modeling, developing a commercial-grade software package has also helped to round out our overall understanding of the formulation of problems we approach in this space. Our brilliant development team includes Justin Richmond, Ryan Kersch, Ryan Hennings, Nancy Hamilton, Jun Sakurai, Matt Cummins, Makoto Scott-Hinkle, Aaron Little, Josh Bauer, Trent Donelson, Tony Ikenouye, Chris Jellings, Justin Holman, and Adam Light. And, thanks to the ILOG CPLEX team, who provided us with their expertise in optimization.

Others who have offered us a good deal of support and a strong voice both out to the market and providing meaningful feedback in return over the years include Chip Wilkins, Aimee Emery-Ortiz, Chris Trivisonno, Bill Huthmacher, and Kitte Knight.

We would like to take the time to extend a special thanks to three members of our team who have had a big influence on all of us and will live on in our memories: Alex Brown, Allen Crowley, and Leslie Smith.

Finally, we would like to thank John Caltagirone from Loyola University and Bill Nickle from Nickle Consulting for reviewing the prepublished book. And last but not least Jeanne Glasser Levine, our editor at Pearson, whose expert guidance and patience with us throughout this process is greatly appreciated.

We appreciate and have enjoyed all the meaningful interactions we have had with so many skilled professionals in this field over the past decade.

ABOUT THE AUTHORS

Michael Watson is currently the world-wide leader for the IBM ILOG Supply Chain Products. These products include the network design product, LogicNet Plus XE. He has been involved with this product since 1998 when the product was owned and produced by LogicTools and then sold to ILOG in 2007 prior to being acquired by IBM. During this time, he has worked on many network design projects, helped other firms develop network design skills, and helped shape the direction of the group and product. He is an adjunct professor at Northwestern University in the McCormick School of Engineering, teaching in the Masters in Engineering Management (MEM) program. He holds an M.S. and Ph.D. from Northwestern University in Industrial Engineering and Management Sciences.

Sara Lewis is currently a world-wide technical leader for the IBM ILOG Supply Chain Products. She has run many full-scale network design studies for companies around the world, she has conducted hundreds of training sessions for many different types of clients, leads a popular network design virtual users group, and helps create educational material for network design. She has been involved with this group since 2006 when the network design tool was owned and produced by LogicTools. Prior to LogicTools, she held various supply chain roles at DuPont. She holds a Bachelor's degree in Business Logistics and Management Information Systems from Penn State University and is a frequent guest lecturer on the topic of network design at several U.S. universities.

Peter Cacioppi is the lead scientist for IBM's network design product, LogicNet Plus XE. He also holds the lead scientist role for IBM's inventory optimization solution. He first began developing network design engines in 1996 as employee number one for LogicTools, a supply chain planning company that was sold to ILOG in 2007 prior to being acquired by IBM. His responsibilities include translating business design issues into formal mathematical problems. His scientific contributions have ranged from developing a targeted network design computational engine to designing both the GUI and the engine for network design multi-objective analysis. He holds an M.S. in Computer Science (with a thesis in Operations Research) from the University of Chicago, and a BA in Computational Physics from Dartmouth College.

Jay Jayaraman currently manages the ILOG Supply Chain and Optimization consulting services team within IBM. This team solves clients' most challenging supply chain and optimization business problems. He brings extensive hands-on expertise in supply chain network design and inventory optimization, with projects ranging from large scale, global supply chain network design strategy to implementing production planning and inventory optimization projects at the tactical level. He has successfully led and managed consulting projects for clients around the world and in many different industries such as chemicals, consumer packaged goods, retail, transportation, pharmaceutical, and many others. Prior to joining LogicTools (later acquired by ILOG and then IBM), he worked for Kuehne & Nagel, helping run network design projects and implement the results. He holds an M.S. in Industrial Engineering from the University of Florida, and a Bachelor's degree in Industrial Engineering from Anna University, India.

PREFACE

This book is aimed at an important and underserved niche within the supply chain educational market: strategic supply chain network design. Almost all supply chain professionals need to know about this discipline, and this book is meant to help them do it well.

Strategic network design is about selecting the right number, location, and size of warehouse and production facilities. At the same time, firms will need to determine the territories of each facility, which product should be made where, and how product should flow through the supply chain.

Typically, these studies do not produce a single correct answer. Instead, strategic network design planning requires you to construct a logical mathematical model of your supply chain, use powerful optimization engines to sort through the seemingly countless possibilities to return the optimal solution, and then analyze the results of many different scenarios to *support* a good operational decision. Fortunately, commercially available software takes care of the calculations for you. However, it is important that you understand what the optimization engine is doing so you can construct better models and deliver better results.

This is an educational book for those who have to do this work and managers who are responsible for an efficient supply chain. Most books in this area are too high level and don't provide enough detail to actually do such work at a firm or to even know whether the principles are being correctly applied. We are going to start with the fundamentals and build on that. Along the way, we will provide you with the practical aspects of the problem, the key insights and intuition from each section, and some background on the theory.

This book covers both the theory and the practice of this discipline in a way that ensures that readers will be successful when implementing strategic network design planning in their respective fields. Doing this work well requires an understanding of the theory. We have included sections on the mathematical formulations of the problems to give you, the reader, a deeper appreciation of how solutions are found. However, since this is a business book, you can skip the math formulations without loss of continuity (but if you skip the equations, we still suggest you read the description to gain additional insight). We will also cover how to get answers in the complexity of the real world. The text is combined with interactive exercises using commercially available software as well. The use of the commercially available software enhances the course and its use may also be skipped without loss of continuity.

This topic is very rich and deep. Several books could be written to cover all the clever ways network design modeling is used in industry and all the different decisions and factors you can consider. Even though we cannot cover every topic and every nuance, this book is meant to introduce you to the topic and give you the foundation to build these more complex models over time. We have found that doing these projects will help build up your skills and insights. This book is meant to get you started in the right direction.

The authors of this book all have extensive experience with these types of projects. We have run hundreds of studies; presented results to managers, CEOs, and boards of directors; and even legally testified on the results to the U.S. Government. We have designed and coded this type of commercial software used by many companies and organizations in many different industries all around the world. We have trained others to do these projects on their own. We have also taught this subject matter in leading universities. All these experiences together have brought us to realize the value of bringing the contents of this book to students and business professionals worldwide.

We want you to be able to learn from our experience.

For those new to the field, we want you to understand and appreciate this interesting discipline.

For those who will need to do this professionally, we want you to be able to do these projects better and with a deeper understanding.

For those who will need to manage these types of projects, we want you to be able to provide better guidance to your team and deliver better results.

For those who teach this material, we want this material to help make your course richer. We want you to have insight into how this gets implemented in practice and gain access to realistically sized problems. You can also use this material to expand into other supply chain areas as appropriate for your course. This book would be appropriate for courses in Supply Chain, Operations Research, Industrial Engineering, or an MBA or Business program. Supply chain network design is about determining the best supply chain to physically move product from the source to the final destination. As such, this topic gives you a great sense of a firm's supply chain and how a firm works. It also touches very strategic business issues such as cost, service, and risk.

For everyone, we hope you find the contents of this book interesting and helpful over the course of your careers.

Book Web Site

This book is written so that it can be valuable as standalone text and enhanced with the ability to do hands-on exercises. By working on the hands-on exercises, you can reinforce the learning from the text and explore the concepts on your own.

The Web site for this book is www.NetworkDesignBook.com.

Information on the exercise materials can be found at www.NetworkDesignBook.com/academic-use/.

In the "Course Material" section, you will find the following:

- Models from the chapters so you can work with the same models we mention in the text

- Excel files with data referenced in the chapters

- Instruction manuals for using the software with the book

- Models and Excel files for the end of chapter exercises

- Additional questions and case studies to keep the material fresh and to allow you to explore new areas

- Updates to the text through time

- Instructor materials, if you are teaching from the book

The other sections of the Web site have additional information, such as links to references, case studies, and recent developments, to enhance the value of the book.

1

THE VALUE OF SUPPLY CHAIN NETWORK DESIGN

What Is Supply Chain Network Design and Why Is It Important?

A firm's supply chain allows it to move product from the source to the final point of consumption. Leading firms around the world, from large retailers to high-tech electronics manufacturers, have learned to use their supply chain as a strategic weapon. A supply chain is defined by the suppliers, plants, warehouses, and flows of products from each product's origin to the final customer. The number and locations of these facilities is a critical factor in the success of any supply chain. In fact, some experts suggest that 80% of the costs of the supply chain are locked in with the location of the facilities and the determination of optimal flows of product between them. (This is similar to the notion from manufacturing that you lock in 80% of the cost to make a product with its design.) The most successful companies recognize this and place significant emphasis on strategic planning by determining the best facility locations and product flows. The discipline used to determine the optimal location and size of facilities and the flow through the facilities is called supply chain network design.

This book covers the discipline of supply chain network design. Sometimes it is referred to as network modeling because you need to build a mathematical model of the supply chain. This model is then solved using optimization techniques and then analyzed to pick the best solution. Specifically, we will focus on modeling the supply chain to determine the optimal location of facilities (warehouses, plants, lines within the plants, and suppliers) and the best flow of products through this facility network structure.

Here are four examples to illustrate the value of supply chain network design.

Example #1

Often, we hear about firms acquiring or merging with another firm in the same industry to reduce the overall costs to operate both firms. That is, they justify the new combined company by determining that they can deliver the same or more products to the market at an overall lower cost. In firms that make or ship a lot of products, a large portion of the savings comes from the merger of the two supply chains. In such mergers, the savings often come from closing redundant plant and warehouse locations, opening new plants and warehouses, or deciding to use existing facilities to make or distribute different mixes of product. We have heard firms claim resultant supply chain savings from $40 million to $350 million over a period of a couple of years. With these kinds of savings, you can only imagine the pressure placed on the supply chain team to determine the new optimal supply chain structure after an acquisition or merger is announced.

Example #2

Often, a large firm will find that its supply chain no longer serves its business needs. In situations like this, the firm will have to transform its supply chain. It may have to close many facilities, open many new ones, and use facilities in a completely different way. For example, a retail firm may have to redesign their supply chain to serve their stores as well as their new online customer base in a more integrated approach. Or a large retailer may find that some of their product lines have grown significantly and the retailer needs new warehouses to manage this growth. If done right, this type of supply chain transformation can help reduce logistics and inventory costs, better respond to different competitive landscapes, and increase sales and profitability. We have even seen firms highlight this work in their annual reports, therefore showing the importance of this analysis to the firm as a whole.

Example #3

In the spring of 2011, we were working on a project for a global chemical company to help develop their long-term plan for their supply chain. This study was analyzing where they should locate new plants to serve a global customer base. The long-term project suddenly became extremely short-term when the CEO called the project team to inform them that within six hours they were closing their plant in Egypt due to political unrest. He also indicated there was no timeline for reopening the plant. The CEO immediately needed to know which of the existing plants should produce the products that were currently being manufactured in Egypt and how customer demand was going to be impacted. The team was quickly able to deliver the answers and minimize this supply chain disruption. As seen in this example, supply chain network design models can also be a great tool for identifying risks and creating contingency plans in both the short and long term.

Example #4

As consumer behavior and buying patterns change, firms often want to bring their product to the market through different channels. For example, we worked with a consumer products company that wanted to analyze different channels such as selling through big-box retailers, selling through smaller retailers, selling direct online, and selling through distributors. This firm wanted to analyze different ways to bring their product to market and understand what the supply chain would need to look like for each of these cases. That is, they wanted to determine the optimal number and location of plants and warehouses. This would then be a key piece of information to help them determine their overall strategy.

Of course, the details of these studies can be a bit more complicated. As an example, take a minute to think through the possible supply chain for a tablet computer and compare that to the supply chain you envision for a candy bar.

The tablet supply chain faces specific challenges surrounding a time-sensitive delivery of the device for a very demanding high-tech customer market. The tablet maker must also determine how to best balance its partnerships with many contract manufacturers worldwide while still ensuring the highest quality end products. Finally, this supply chain must deal with the high costs for insurance, transport, and storage of these high-priced finished goods.

Conversely, when we shift our thoughts to the supply chain related to the candy bar, we must consider an entirely different set of challenges and objectives that the candy maker must face: government regulations that mandate different requirements for all stages of production paired with a strict shelf life of each unit produced. In addition, raw material costs, as well as costs tied to temperature control during transit and storage, add up. Major swings in demand due to seasonality or promotions also add the need for flexibility within their supply chain.

Despite their differences, both the tablet maker and candy bar maker must determine the best number and location of their suppliers, plants, and warehouses and how to best flow product through the facilities. And building a model using optimization is still the best way for both of them to determine their network design.

As the previous examples highlight, many different types of firms could benefit from network design and many factors go into the good design of the supply chain. Along with ever-growing complexity, the need to truly understand how all these requirements affect a company's costs and performance is now a requirement. Using all these variables to prove out the optimal design configuration commonly saves companies millions of dollars each year.

As you would expect, a network design project can answer many types of questions such as these:

- How many warehouses should we have, where should they be, how large should they be, what products will they distribute and how will we serve our different types of customers?

- How many plants or manufacturing sites should we have, where should they be, how large should they be, how many production lines should we have and what products should they make, and which warehouses should they service?

- Which products should we make internally and which should we source from outside firms?

- If we source from outside firms, which suppliers should we use?

- What is the trade-off between the number of facilities and overall costs?

- What is the trade-off between the number of facilities and the service level? How much does it cost to improve the service level?

- What is the impact of changes in demand, labor cost, and commodity pricing on the network?

- When should we make product to best manage and plan for seasonality in the business?

- How do we ensure the proper capacity and flexibility within the network? To meet demand growth, do we need to expand our existing plants or build new plants? When do we need to add this capacity?

- How can we reduce the overall supply chain costs?

Being able to answer these questions in the optimal manner is important to the overall efficiency and effectiveness of any firm. Companies that have not evaluated their supply chain in several years or those that have a new supply chain through acquisitions can expect to reduce long-term transportation, warehousing, and other supply chain costs from 5% to 15%. Many of these firms also see an improvement in their service level and ability to meet the strategic direction of their company.

Although firms are happy to find 5% to 15% reduction in cost, it does highlight that your supply chain might have already missed out on significant savings you may have realized had you done the study a year ago (or two or more years ago). Some firms have realized this and now run this type of analysis on a more frequent basis (say, quarterly). This allows them to readjust their supply chain over time and keep their supply chain continually running in an optimal state while preventing costs from drifting upward.

The frequency of these studies depends on several factors. Historically, it has been customary to complete these analyses once every several years per business unit, because it was usual for business demographics and characteristics to change over this period of time. For some industries such as high-tech, the frequency was even higher because there may be higher volatility in customer demand, thereby requiring periodic reevaluation of the network. Any major events, such as mergers, acquisitions, or divestitures, should also trigger a network reevaluation study. As noted before, the savings from the optimization of the revised network typically represent a significant part of the savings that justify the merger or divestiture. A current trend we are seeing, however, is to do these studies even more frequently. Business demographics and characteristics are changing faster. In addition, the growth of the global supply chain is driving firms to cycle through studies as they go from region to region around the world. Also, firms are running the same models more frequently to stay on top of changes in their business by adjusting the supply chain. Some firms update these models several times throughout the year.

Determining the right supply chain design involves a lot of quantitative data as well as some nonquantitative considerations. We will discuss this in the rest of this chapter, as well as how we use mathematical optimization to sort through this quantitative data.

Quantitative Data: Why Does Geography Matter?

It should be clear by now that the supply chain network design problem is just as much about geography as it is about business strategy. The two cannot be separated.

Take these supply chain considerations for example:

- If you have a plant in the interior of China and some of your customers are in New York, you need to physically get the product out of China, across the ocean, and to New York.

- If you make wood products (like paper or boards), you can locate plants either close to the raw materials (forest areas) or close to your customers (usually located a significant distance away from the large forest areas).

- If you have a warehouse in Indianapolis, you are close to your customers in Chicago, but far away from customers in Miami. If the warehouse is in Atlanta, you are closer to Miami, but farther from Chicago.

- If you make a critical product only in Miami, a hurricane may shut down your operation, causing a loss of revenue.

As the examples highlight, decisions about the location of your facilities impact many aspects of your business and require you to make trade-offs. Specifically, geography drives the following:

- **Transportation Cost**—You need to move product from its original source to its final destination. The location of your facilities determines the distance you need to move product, which directly impacts the amount you spend on transportation. But, also, the location of your facilities determines your access to transportation infrastructure such as highways, airports, railheads, and ports. Finally, because of supply and demand, different locations may have different transportation rates.

- **Service Level**—Where you locate relative to your customers impacts the time it takes to get product to your customers. For some products, you can negate great distances by using overnight air freight. But this usually comes at a premium cost.

- **Risk**—The number and location of your facilities impacts risk. If you have just one location for a critical activity, there is always the risk that a fire, flood, some other natural disaster, a strike, or legal issues will shut down your operation. There is also political risk to consider. Your facility could get confiscated or shut down for political reasons, or the borders may shut down, isolating your facility.

- **Local Labor, Skills, Materials, and Utilities**—The location of your facilities also determines what you pay for labor, your ability to find the needed skills, the cost of locally procured materials (which is often directly related to the local labor costs), and the cost of your utilities.

- **Taxes**—Your facilities may be directly taxed depending on where they are located and the type of operations being performed. In addition, you also need to consider the tax implication of shipping product to and from your locations. In some industries, taxes are more expensive than transportation costs.

- **Carbon Emissions**—Locating facilities to minimize the distance traveled or the transportation costs often has the side benefit of reducing carbon emissions. In addition, if your facilities consume a lot of electricity, you can reduce your emissions by locating near low-emission power plants.

As the list highlights, geography matters. What makes this challenging is that the geography often pushes the solution in different directions at the same time. For example, it would be desirable to have a facility close to all the demand. However, demand is typically where people live. And it is usually very expensive, if not impossible, to locate a plant or warehouse in the middle of a major metropolitan area. So the desire to be close to customers pushes locations close to cities. The desire for cheap land and labor (and welcoming neighbors) pushes the best locations further from the city center. In global supply chains these decisions become even more extreme. In some cases it may make sense to service demand from a location on an entirely different continent.

In addition to geography, the next two sections will discuss the importance of warehouses and multiple plants to your supply chain as well.

Quantitative Data: Why Have Warehouses?

In this book, a warehouse represents a facility where firms store product or a location where product simply passes through from one vehicle to another. It can be called a distribution center, a mixing center, a cross dock, a plant-attached warehouse, a forward warehouse, a hub or central warehouse, a spoke or regional warehouse, or a host of other terms.

To understand how to optimally locate warehouses, it is important to discuss why warehouses exist. Wouldn't it be much cheaper for companies to load the product only once, at the manufacturing location, and ship it directly to the customer? Stopping at a warehouse adds loading, unloading, and storage costs, not to mention the cost for two legs to transportation (one leg from the plant to the warehouse and one leg from the warehouse to the customer). In cases where you can ship directly from the plant, it is usually good to do so. Therefore, it is important to ask questions to see whether you can avoid warehouses altogether. But in most cases warehouses are needed in a supply chain for the following reasons:

- **Consolidation of Products**—Often, you will need to deliver a mix of different products to your customers and these products may come from various sources. A warehouse serves the useful function of bringing these products together so that you can then make a single shipment to a customer. This will be cheaper than having the products ship to the customers directly from each individual source of supply.

- **Buffer Lead Time**—In many cases, you will need to ship to your customers with lead time that is shorter than that which can be offered by shipping directly from the plant or supplier location. For example, you may promise to ship products to your customers the next day but your plants or suppliers may have a lead time of several weeks before they are able to make the product available to the customer. In this case, the warehouse holds product at a location closer to the customers in order to provide the next day transport promised each time an order is placed.

- **Service Levels**—Where you store the product and its proximity to the market where it will be consumed is also a measure of the service level the company can provide. The need to be close to customers can create the need for multiple warehouses. Overall cost versus service level is one of the most classic trade-offs in supply chain network design.

- **Production Lot Sizes**—Setting up and starting the production of a single product or group of similar products on a line can have a significant fixed cost associated with it. Therefore, production plans attempt to maximize the number of units of product made during each run. (This production amount is called a lot size.) Understandably, these lot sizes normally do not match the exact demand from the market at the time. This requires the extra units to be "stored" in warehouses until future demand requires them. Production lot sizes versus inventory storage costs is also a common supply chain design trade-off.

- **Inventory Pre-Build**—Some industries see huge spikes in the supply of raw materials (seasonal food harvests) or in the demand of finished goods (holiday retail shopping). In the case of raw material supply spikes, some firms must store these abundant raw materials until the time they will be needed for steady monthly production cycles. Other firms must immediately use these raw materials to produce finished goods that are not yet demanded. These additional finished goods must then be stored until demand in future time periods requires them. In the case of demand spikes, companies find themselves with insufficient production capacity to fulfill all orders during peak periods of demand. As a result, they must use their additional capacity during off-peak time periods to make finished-good units to be stored awaiting their use to fulfill the upcoming spikes in demand. The use of costly overtime production versus inventory storage costs is another common supply chain design trade-off.

- **Transportation Mode Trade-offs**—Having warehouses often allows you to take advantage of economies of scale in transportation. A warehouse can help reduce costs by allowing the shipment of products a long distance with an efficient (and lower cost) mode of transportation and then facilitating the changeover to a less efficient (and usually more expensive) mode of transportation for a shorter trip to the final destination (as opposed to shipping the entire distance on the less efficient mode).

It is also important to match up the preceding list of reasons for warehouses with the types of warehouses in the supply chain. A supply chain may have many types of warehouses to meet many different needs. Here are some common types of warehouses:

- **Distribution Center**—Typically refers to a warehouse where product is stored and from which customer orders are fulfilled. This is the most common and traditional definition of a warehouse. When a customer places an order, the distribution center will pick the items from their inventory and ship them to the customers. These types of facilities are also called *mixing centers* because they "mix" products from many locations so that your customers can place and receive an order from a single location. If a manufacturing company does not have this type of warehouse in the supply chain, customers may have to place several orders or receive several shipments from different locations depending on where each product they want is made.

- **Cross-Dock**—Usually refers to a warehouse that is simply a meeting place for products to move from inbound trucks to outbound trucks. The term simply means that products pass (or cross) from one loading dock (for inbound trucks) to another loading dock (for the outbound trucks). For example, in the case of a produce retailer with 50 stores, they may have a full truck of fresh peaches arriving at the inbound docks from a single supplier. The peaches are then removed from the truck and some are placed in each of the 50 waiting trucks on the outbound side, according to the relevant store demand. This happens for peaches as well as a host of other produce items. Basically full trucks arrive from a single supplier on the inbound side of the facility, and then transferred to multiple trucks on the outbound side of the facility resulting in fully loaded truckloads with a mix of product from each of the suppliers quickly sent on their way. The best-run cross-dock systems have all the inbound trucks arriving at approximately the same time so that product stays at the cross-dock for only a short period of time.

- **Plant-Attached Warehouse**—Refers to a warehouse that is attached to a manufacturing plant. Almost all plants have some sort of product storage as part of their operations. For some, it may simply be a small space at the end of the line where product is staged prior to being loaded onto a truck for shipment. In other cases, the warehouse can act as a storage point for product made at the plant or for products made at other plants. In this case, this warehouse acts like a distribution center co-located with the manufacturing facility. A major benefit of a plant-attached warehouse is the reduction of transportation costs because a product does not have to be shipped to another location immediately after it comes off the end of the line. When you have plant-attached warehouses, sometimes the standalone warehouses are called *forward warehouses,* meaning they are placed "forward" or out closer to customers.

- **Hub Warehouse or Central Warehouse**—Refers to a warehouse that consolidates products to be shipped to other warehouses in the system before moving on to customers. Different from cross-docks, the products are normally stored in these locations for longer periods of time before being used to fulfill demand. The other warehouses in the network are then typically called *spokes* or *regional warehouses.*

In practice, you will find many different names for warehouses. These names are most likely just different terms for what is described in the preceding list. In addition to the types of facilities, there are also needs for different temperature classes (frozen, refrigerated, or ambient), different levels of safety (hazardous or nonhazardous), and different levels of ownership (company owned, company leased, or the company uses a third-party facility).

As an interesting side note and to further illustrate the wide range of warehouse types we have experienced, we have even seen caves used as warehouses. Caves have the nice advantage of maintaining the same (relatively low) temperature and have prebuilt roofs.

If you can get trucks into and out of them and have room to store products, caves also make great warehouses. Kansas City is probably the best-known location with cave warehouses.

Quantitative Data: Why Have Multiple Plants?

Similar to our use of the word "warehouse," we are also going to use the term "plant" to broadly refer to locations where product is made or where it comes from. So a plant could be a manufacturing plant that produces raw material, components, or finished goods, or just does assembly. A plant could be owned by the firm, it could be a supplier, or it could be a third-party plant that makes products on behalf of the firm. These third-party plants are often called co-packers, co-manufacturers, or toll manufacturers (a term used for third-party manufacturers common in the chemical industry).

A plant can also contain multiple production lines. So, often we are determining not only the location of plants, but also the number and location of the production lines.

For companies that make products, the number and location of plants are important. For retailers and wholesalers however, the location of the suppliers is often beyond their scope of control.

Even for a firm that has only one product, the location of the plant can impact transportation costs and the ability to service customers. In some cases, the location of the plant is primarily driven by the need to have skills in the right place or the need to be next to the corporate headquarters. However, in most cases, there are choices for plant locations. An even more interesting choice is to determine whether you should locate multiple plants to make the same product—even when a single plant could easily handle all the demand.

When you have choices for where to put your plants or the option to have several plants, you must consider some of the same questions we did when locating warehouses. For example, factors that would drive you to have multiple plants making the same product include:

- **Service Levels**—If you need your plants to be close to customers, this will drive the need for multiple plants making the same product. This becomes especially important if your business does not use warehouses. In this case, your plants face the customer and their location can drive service levels.

- **Transportation Costs**—For producers of heavy or bulky products that easily fill up truckload capacities, you will want to be as close to your markets as possible. This may also drive the need for multiple warehouses.

- **Economies of Scale**—As a counterbalance to the benefits of transportation, you also want to factor in the economies of scale within production. As mentioned previously, the more you make of a given product at a single location, the lower the production cost per unit. This is driven by a reduction in production line setup time and costs and the benefit of being able to create a more focused manufacturing process. So while it may be ideal to have many production locations to minimize transportation cost, economies of scale in manufacturing suggest that fewer plants will be better.

- **Taxes**—In a global supply chain, it is often important to consider the tax implications of producing and distributing product from multiple or different locations.

- **Steps in the Production Process**—In a production process with multiple steps, you may need to decide where you should do various activities. For example, it can often be a good strategy to make product in bulk at a low-cost plant, ship it in bulk to another plant closer to the market to complete the conversion to a finished good.

As with the warehouses, a plant can also represent many types of facilities. A plant may represent any of the following:

- **A Manufacturing or Assembly Site**—This is a site that is owned by the firm that makes products. The products coming out of the plant could be raw materials, semifinished goods, or finished goods.

- **A Supplier**—This is a location that is not owned by the firm but supplies product to the firm. The products could again be raw material, semifinished goods, or finished goods. In the majority of cases, you have no control over where your suppliers are located. You may, however, be able to pick which supplier you will purchase from.

- **Third-Party Manufacturing Site**—This is a location similar to supplier plants, but these sites make product on behalf of the firm and are therefore treated more like the firms' own plants than a raw material supplier. Many firms use third-party manufacturing sites because manufacturing may not be their core competency. Their competitive advantage comes from all other activities including the design, marketing, and/or final sale of the product. These third-party firms, typically called *contract manufacturers,* are widely used in the electronics industry. Other third-party firms, termed *co-manufacturers, co-packers,* or *toll manufacturers,* are used to simply supplement the capacity of the firm itself. These are quite common with consumer packaged goods like food and beverages or chemical companies.

As with warehouses, in practice, you will find many variations in terminology for plants and different types of manufacturing sites and suppliers serving roles similar to those mentioned previously.

Solving the Quantitative Aspects of the Problem Using Optimization

Because of the supply chain complexities and rich set of quantitative data we have discussed, mathematical optimization technology is the best way to sort through the various options, balance the trade-offs, determine the best locations for facilities, and support better decision making. The mathematical optimization relies on linear and integer programming.

One common misconception we see is that managers sometimes confuse having good data and a good reporting system with having an optimal supply chain. That is, they think that their investment in good data and reporting systems should equip them with the ability to complete network design analysis. (These systems are often called business intelligence systems.) But in reality, if your warehouses or factories are located in the wrong places, these reporting systems will not correct the situation or suggest new locations.

So although these systems do, in fact, generate good reports and allow managers to gain good insights, they do not lead to better designs for the supply chain. That is, they are not built to construct models of your supply chains to which you can apply mathematical optimization. At best, they may allow you to evaluate a small handful of alternatives, but in doing so, you have to define all the details of each of the alternatives. Optimization is a complementary, not competitive, technology that allows you to actually determine the best locations for your facilities. And you can let optimization do the heavy number crunching to determine the details of the alternatives (where to locate, what is made where, how product flows, which customer is served by which warehouse). And, in many cases, if the optimization is set up correctly, it will uncover ideas that you never thought about.

Of course, because these problems are of great strategic value to an organization and touch on many aspects of the business, there will be nonquantitative aspects you must consider. These nonquantifiable aspects are important and are discussed later in this chapter and throughout the book.

With all that said, solving the quantitative aspects of this problem with mathematical optimization is the key to coming up with the best answers. So, let's start there.

To formulate a logical supply chain network model, you need to think about the following four elements:

- Objective
- Constraints
- Decisions
- Data

The *objective* is the goal of the optimization and the criteria we'll use to compare different solutions. For example, the most common objective in strategic network design is to minimize cost. If our objective is to minimize cost, we can now compare two solutions and judge which one is better based on the cost. When the mathematical optimization engine is running, it searches for the solution with the lowest cost. So with this common example you can see that an optimization problem needs to have a quantifiable objective. It is important to point this out because we have encountered many situations in which supply chain managers say their objective is to "optimize their supply chain." The appropriate response is to ask what exactly they want to optimize, or what criteria they will use to determine which of two solutions is better. For example, is the key criteria transportation cost, is it service, is it facilities costs, or something else? Later, we'll discuss optimizing multiple objectives (as long as you can quantify them). And we will show methods for analyzing nonquantifiable factors such as risk and robustness, because these are also very important. But for now, we'll start with one quantifiable objective. When we understand this one, we'll be able to understand more detailed analysis later.

The *constraints* define the rules of a legitimate solution. For example, if you want to just minimize cost, it is probably best to not make any products, not ship anything, or have no facilities. A cost of zero is indeed the minimum; however, that's clearly not realistic. So there are some logical constraints we must include such as the fact that you want to meet all the demand. There are also constraints that specify which products may be made where, how much production capacity is available, how close your warehouses must be to customers, and a variety of other details. In this step, you also want to be careful not to specify so many constraints that you prohibit the optimization from finding new and creative strategies.

The *decisions* (sometimes called decision variables) define what you allow the optimization to choose from. So in the optimization of the physical supply chain, the main decisions include how much product moves from one location to another, how many sites are picked, where those sites are, and what product is made in which location. But, certainly, there are other decisions as well. The allowable decisions cannot be separated from the constraints. For example, if you have existing warehouses, you may or may not be able to close some of these sites.

Finally, you must consider the *data* you have available to you. There may be factors you would like to consider in the optimization, but you do not have the data to support. In this case, you still need to figure out ways to make a good decision. This could include running multiple scenarios, considering approximate data, or adjusting the data you have. We will further discuss a variety of these techniques later in this text.

After you have thought about your problem in terms of the objective, constraints, decisions, and data, there are ways to translate this into a series of equations and then solve the problem using linear and integer programming techniques. This is also sometimes referred to as mixed integer programming (MIP)—the "mix" refers to a mix of linear and integer variables. We will cover what this means in more detail later in the book, and will also provide information on how these problems work. There are whole courses on MIP however, so we will cover the topic only as it relates to supply chain network design problems in this text.

One way to think about a MIP is to think of it as a series of steps that are influenced by the objective, the constraints, and the decision variables. That is, during the steps, the objective steers the solution to more favorable costs and avoids less favorable costs, the constraints set the rules and can prevent it from doing more of what it wants or can force it to do something that is not favorable to the objective, and the decision variables tell it what it is allowed to change. A nice benefit of MIPs is that they solve for all the decisions and consider all the constraints simultaneously. That is, this approach allows you to come up with the overall best solution for a given problem.

Data Precision Versus Significance: What Is the Right Level in Modeling?

The discussion about quantitative data brings us to a discussion about data accuracy.

The lessons we learned about significant digits in high-school chemistry will serve us well when doing network design studies.

We learned that measurements were only as accurate as the equipment that took the measurement, and we had to report it as such. And when we combined different measurements together in some equation, our answer could be reported only in terms of the measurement with the least amount of accuracy. This accuracy was expressed in the number of significant digits. For example, if we took one measurement and it was 3.2 units and added that to another measurement of 4.1578 units and added them together, our calculator would give us 7.3578 units. However, we have to report it as 7.4 units because the final answer can have at most two significant digits. To write it otherwise would give the answer a level of accuracy that just wasn't true.

We often assume that more precise data is always better. However, as our lessons in significant digits taught us, we can be only as precise as our measurements allow. Keeping this concept in mind will serve you well when you are collecting data for your network design models as well.

The adage about bad data leading to bad results, "garbage in equals garbage out," can sometimes just confuse. This adage does not mean that data needs to be precisely measured to a certain number of significant digits. It just means that the data has to be good enough for the decisions we are making.

Other problems can result when the precision and detail of the data actually get in the way of making good decisions. The cartoon in Figure 1.1 highlights this point very well.

Figure 1.1   Precision Cartoon

In the cartoon, the extra precision on the left actually makes things worse for our poor analyst (who is about to be hit by a piano). The analyst has to spend too much time trying to understand the data and misses the opportunity to take the much-needed action of getting out of the way. In network modeling, our time horizon for making decisions is much longer, but the data is also much more complex. We have seen projects in which the extremely detailed analysis of data causes the project team to miss their opportunity to impact the supply chain in a positive way.

Our goal for collecting data for a network design model is to define the data needed with the right level of significance to make the relevant decisions. Our ultimate responsibility is to report the results with the right level of significance for the organization to make decisions. Our goal is not to ensure that every piece of data is significant to ten decimal places. Therefore it is really a waste of time and often a risk to our project's success when we report data with more significance than is warranted.

Also keep in mind that when we are running a network design project, we are making decisions about the location of our plants and warehouses to support the business in the *future.* It is pointless to build a new warehouse to support last year's business. Although you may start your analysis with historical data as a baseline (we'll cover baseline in more detail in Chapter 8, "Baselines and Optimal Baselines"), you will eventually want to consider how this data might be different when considering it in a future state. Two specific elements with future uncertainty, for example, include:

- **Demand Data**—There will be uncertainty in future demand dependent on overall economic conditions, moves by competitors, success of your marketing programs, and so on.

- **Transportation Costs**—There can be a lot of variance in future prices of transportation dependent on the ever-fluctuating world market price of oil.

Based on the previous, we know our instruments for measuring future demand and transportation cost data are likely going to be inaccurate. Therefore we should consider using fewer significant digits when representing them in the model. If we get push-back to include more significant digits from other stakeholders in the project, we should remind ourselves that if we could accurately predict future demand or the future price of oil, we would benefit more by taking this knowledge to the financial markets.

In Chapter 4, "Alternative Service Levels and Sensitivity Analysis," we will expand our discussion of this topic to include the use of sensitivity analysis and multiple scenarios to help make our best network design decisions despite this problem of uncertainty. That is, just because we don't know the future demand or the future price of oil, doesn't mean we can't still come up with good solutions.

Often times, the problems we mentioned previously tempt teams into giving up on network design until they get better and more precise data. People get nervous about making decisions without enough significant digits or enough precision. There are two problems with this approach, though:

- You are fooling yourself that you are not making decisions. If you don't do anything with your supply chain, you are making the decision that there is nothing you can do to improve the supply chain right now. And if you make decisions without any formal data collection and modeling, you are using the resource that is the most imprecise and has the fewest significant digits of all: your intuition.

- You are missing a chance to better understand your supply chain and understand what data you should be collecting. For example, a good practice that we have found is to make initial assumptions, run scenarios, and test those assumptions by varying the data by +/−10%, then +/−20%, and seeing what happens. These runs give you insight into your supply chain and show you the value of the data. For example, if the results are not sensitive to a particular data element, you do not have to spend much time refining that data element as you continue to build out your model.

Of course, when we are looking at the results and output data, we also want to keep our lessons in significant digits in mind. If our significant digits are valid only to the nearest million dollars, then a reported savings of $500,000 should not be considered.

The other concept to keep in mind when analyzing the output comes from introductory statistics. When you first learned about hypothesis testing, you were shown techniques for determining whether two different samples were statistically different from each other. For example, if you ran two marketing campaigns in two different markets and one market came back with average sales of $15,000 per store and the other with $17,500 per store, you could not immediately conclude that the second marketing campaign was better. In general, the higher the variability of the data in the sample size, the harder it is to claim that there is a statistical difference between the two samples.

Although we cannot apply this direct statistical test to our network design models, we can apply the concept. Because we already know that we have a lot of underlying variability in our data (such as future demand, future costs, and so on), we know that we need to show a fairly large savings from the current state to be comfortable with making the decision to implement the recommended change.

Typically, we look for savings of more than 5% to 10% compared to the current situation before we recommend a change. That is, when relocating facilities, if you find savings that reduce costs by 1% to 2% or change services levels by 2% to 3%, we would consider this not statistically different from the current state. When the savings are 15% to 25% however, you can be confident that you have found a statistically different solution.

Of course, you should be careful with this rule. You don't need to completely disregard the value of relatively small improvements either. The key is to prevent the magnitude of the suggested change from swamping the size of the expected benefit. For example, you might study a $300 million supply chain that is already well rationalized, and only find $250,000 in savings. However, this 0.084% savings might be well worth the bother, if it could be realized by reassigning just two customers. In such a situation, you might think of the $250,000 as being proportionally quite large when compared to the total landed cost of serving these two wayward demand points.

As you go forward in this book and with projects for your company, these are important lessons that will serve you well.

Nonquantifiable Data: What Other Factors Need to Be Considered?

Our discussion up to this point has focused on the quantifiable aspects of the network design solution. However, there are other factors that you want to consider when making a final decision. Some of these factors do not lend themselves to being quantified and being considered directly within the optimization runs. This does not mean that our optimization runs are meaningless, nor does it mean that we should just ignore these factors.

It is important to run the optimization with as much quantifiable information as needed. You will want to run a variety of scenarios with different input data. This will then give you a range of potential solutions. From this range, you will then want to apply the consideration of the nonquantifiable aspects. For example, if closing existing facilities and opening new ones in new locations increases your political risk, you want to know whether that new configuration saves you $500,000 or $50 million. You will then be able to judge whether the extra risk is worth it.

We will discuss how you run different scenarios and create a range of possible solutions in future chapters. Some of the nonquantifiable factors you might then want to consider include the following:

- **Firm's Strategy**—Your firm may value cost more than service or vice versa. For example, your firm may have a strategy of servicing the top customers at any expense or be committed to a local manufacturing strategy.

- **Risk**—For global supply chains, you need to worry about placing sites in politically unstable locations, port closures, and the added risk of extra distance between origins and destinations. There is also a risk when you have just a single location to make a given product or you have a supply chain that is currently at capacity and is not equipped to handle any unexpected extra demand.

- **Disruption Cost**—Firms realize that changes could cause significant disruption, leading to other costs like attrition, lost productivity, and unmet demand.

- **Willingness to Change**—Some firms may be more willing to change than others. This can impact the range of solutions you might want to implement. For example, for a firm that is not willing to change, they may be happy to give up savings in exchange for a minimal number of changes.

- **Public Relations and Branding**—This is especially important for firms with a highly visible brand. If one of these firms opens or closes a new facility (especially a manufacturing location), it can often make the news. These firms need to consider the public reaction and the impact on their brand.

- **Competitors**—A firm's supply chain can be impacted by the competition. Sometimes it makes sense to be exactly where the competition is, and in other cases it makes sense to be where they are not.

- **Union versus Non-Union**—Some firms have strong policies on union affiliation or union contracts and want the locations chosen to reflect that.

- **Tax Rebates**—Although taxes can be modeled directly as a cost (as product crosses borders or tax jurisdictions), there can also be rebates for locating a facility in a particular location. This can be hard to quantify during the analysis, but it can be used for negotiating with the local tax authorities.

- **Relationships with Trucking Companies, Warehousing Companies, and Other Supply Chain Partners**—You may have supply chain partners (like trucking or warehouse providers) that you will not be able to work with in new locations. There is some value to keeping your existing partner relationships. You will need to consider who your new partners will be in the new configuration, or what you will need to do to get these new partners.

There are many more such factors. What is important to remember is that we are not just pushing the "run" button and coming up with the right answer. We want to run a variety of scenarios and then apply other criteria when finalizing the decision. When we finalize this decision, we can realize exactly how the quantifiable factors (cost and service) are impacted. This can help lead to discussions based on facts, data, and trade-offs (rather than gut-feel, intuition, and emotion).

Nonquantifiable Data: What Are the Organizational Challenges?

A supply chain study must span many different areas of an organization: sales, operations, logistics, finance, and IT.

The first challenge, aside from merely gathering all these people into the same room at the same time, is to understand and start to balance the different objectives that each group may have. As you can imagine, each of these groups operates with its own specific goals and these may directly conflict with each other. There are many examples of the various groups' goals, so here are just a few related to what we'll cover in this book:

- **Sales Team**—Place product as close to customers as possible (create many warehouses). Have small frequent shipments to customers (many small shipments or more frequent production runs at the plants).

- **Operations Team (Production)**—Produce large quantities of one product during each run in order to reduce machine downtime and changeover costs (creates a need for a lot of warehouse storage). Produce product in one location to maximize economies of scale.

- **Operations Team (Warehousing)**—Quickly move inventory through the warehouses (minimize storage costs). Minimize warehousing locations to reduce fixed and management costs.

- **Logistics Team (Transportation)**—Have large shipments on less costly modes of transportation (ocean, rail, or truckload).

- **Finance Team**—Have the least amount of money tied up in capital (low levels of inventory and operations requiring the least investment in warehousing and production locations). Incur the lowest costs tied to logistics (transportation, warehousing).

Understanding the different objectives of the different groups is important to any successful project.

The second challenge you have is collecting and validating data from all these different parts of the organization. The sales group must produce the appropriate historical demand data as well as dependable forecasts of sales in the future. The operations group will be needed to explain the costs, capabilities, and capacities of all the production and storage assets, as well as any related overhead and labor costs. The logistics group is also needed to provide not only current transportation rates but estimates of rates for new potential lanes resulting from a reorganized network. The finance department is depended on for comparing the output costs from the model to the costs within their financial statements for the same span of time. Doing this provides a validated starting point for the model and a baseline to which we can compare all future model scenarios and output. This data may lie in different systems as well, which only adds to your challenges and often requires IT help to sort out.

Data challenges also come when you are attempting to estimate data for the new potential locations and product flow paths. (Even though this data may be difficult to collect, this is often the whole point of a study—to consider new alternatives.) Transportation rates for new lanes, potential site costs, capacity, and capabilities, as well as the cost to shut down existing sites, must all be researched and calculated for consideration when we ask the model to make the best decision.

The third and final challenge comes after the modeling is done and you have come to the final decision. The final step of actually implementing the results can be a major challenge in and of itself. People in any company become very comfortable with a certain way of doing things. As a result, it is not always easy to get them to see the "big picture" and the value these changes will bring. Proper involvement from all of the previously mentioned teams within an organization throughout the entire project can assist with this, however, as each team understands the rules and constraints that they ensured were adhered to within the recommended solution. There are many great resources for you to learn about how to implement change like this in an organization, but this topic is beyond the scope of this book.

Making changes to a supply chain may also cause a temporary state of disruption. A supply chain cannot just stop at a moment in time and take on a new structure. It is often important to implement changes over a period of time to minimize the downtime and inconvenience that switching over operations may cause.

Despite all of this, however, the more network design projects that companies complete, the better they get at addressing these challenges. And, their future models and recommendations improve as a result. Based on the savings we see from firms that develop this capability, we can say with surety that it is well worth it!

Where Are We Going with the Book?

This book is organized to give you a set of building blocks to tackle any type of supply chain. We will start with very simple problems and build from there. This field is surprisingly deep however, and this book cannot cover every type of model to address every type of business that can benefit from modeling. We have found that advances in optimization technology and computing power have opened up new opportunities for answering different questions and using this technology for more and more decisions and we expect this to continue at an even faster pace in the years to come.

In fact, in just the past ten years, we have seen a dramatic increase in the complexity of models that people want to run and the frequency with which they want to run these models. This is a very positive trend as firms derive more and more value from these types of models.

To successfully deploy the more complex models and run them on a more frequent basis, we think it is critical that you understand the basic building blocks.

You will gain a lot of depth and insight from even the simplest problems. Then, later, as you tackle larger and more complex problems, the foundations you learned from the simple problems will continue to serve you well.

End-of-Chapter Questions

1. ABC Bottling Company's sales have been expanding rapidly. Their single plant, which ships directly to customers, is now out of capacity. What factors should they consider when they decide whether to expand the existing plant or build another one? If they build another plant, what factors should they consider when they locate this plant?

2. A producer of dog food is trying to decide whether they should change the number and locations of their warehouses to better meet projected demand over the next three years. They do a study and determine that their transportation and warehousing costs will be $51 million if they stick with their current structure. They have determined that if they close two warehouses and open two new warehouses, their costs will drop to $50.5 million. Assume that all other costs stay the same. Should they make the change?

3. You need to set up a mathematical optimization model. Assume you are modeling a supply chain for a business with ten warehouses and 1,000 customers. If you set up the model to minimize cost, set the decision variables to decide which warehouse should serve which customers, and set up *no* constraints, why would you expect the minimal cost to come back as $0?

4. You are helping a firm determine their future transportation costs between their plant in Dallas and their warehouse in Atlanta. Your best estimate, with the data you have, is that the cost will be between $1.70 and $1.80 per mile. You decide to use $1.75 as your cost because it is the mid point. If you are asked to spend more time seeing whether the number should be closer to $1.70 or $1.80, what would be your argument against further refining this number?

5. A small medical supply company in Australia has just developed a never before seen product with major pre-release orders from around the globe already. This company will need more production capacity to support their forecasted sales for this new blockbuster product. If they simply expand their plant in Australia, they estimate that their production, transportation, and warehousing costs will be approximately $450 million (AUD). After a careful network design study, they have found two solutions that people in the company generally like.

   a. Solution #1: Estimated cost of $375 million with a new large plant in China to supplement their existing plant in Australia.

   b. Solution #2: Estimated cost of $385 million with three new smaller plants in China, Brazil, and Italy to supplement their plant in Australia. These plants would service their local regions.

   (Assume the costs listed here include all the costs that are relevant.) What would be the best reasons for picking solution #1? For picking solution #2? Why is it important for this firm to consider other nonquantifiable factors when determining their best course of action for expansion?

2

INTUITION BUILDING WITH CENTER OF GRAVITY MODELS

The simplest facility-location problem is the center of gravity (COG) problem. This is also a good place to start building intuition for more complicated models.

Center of gravity problems are ubiquitous in everyday life. When two children position themselves on a teeter-totter so that their different weights are in equilibrium, they are intuitively solving a two-body center of gravity problem. When a waitress loads a tray with an impressively large number of drinks and plates, and then balances the entire arrangement by positioning her hand in an unexpectedly off-center position, she is solving a center of gravity problem.

These problems come from our intuition of physics, and it is natural that the location of facilities in a supply chain borrows the terminology.

For logistics, a center of gravity problem is usually defined as selecting the location of a facility so that the weighted-average distance to all the demand points is minimized. So, in effect, the problem is similar to the waitress balancing the tray. The items on the tray are like the demand points, and the placement of her hand is like the facility location.

In the world of logistics, center of gravity problems are valued precisely because of their simplicity. A center of gravity solution suggests that facilities are located at the center (the "center of gravity") of a collection of demand points (or in some instances, for firms with many suppliers, at the center of the supply points). Another way to think about this problem is to imagine a facility plopped down anywhere on the map. The demand points then engage in a tug of war to pull the facility closer. The larger demand points have more pull. If many small demand points are in a region, they will pull the facility closer to the region. The equilibrium point is when no demand point can pull the facility any closer without creating a solution that is worse for the entire system.

Center of gravity models are, by definition, clear-cut and not ambiguous. Problem formulations do not require a skilled professional to determine the cleanest approximation or modeling formula, but rather, merely require a correct and accurate specification.

Thus, the same input data will inevitably yield the same result (or, in some cases, the same set of distinct, but functionally equivalent results). Center of gravity studies are useful both for building the intuition of the professional analyst and for validating the accuracy of his or her more comprehensive results.

Center of gravity studies are also powerful learning problems for someone new to the mathematical language of optimization. Because these problems are relatively simple, we can show how formal equations are designed to force a numerical engine to generate the optimal solution to the correct problem. In your career, you might be called upon to generate optimization formulas yourself. It is more likely that you will be required to use both off-the-shelf and customized software solutions that implement such formulations automatically. Regardless, it will be useful to have seen how a problem metamorphoses from a real-world "story problem" to a partially formulated collection of mathematical equations, and finally to a carefully tuned set of constraints, variables, and objective function.

By contrast, capturing the fiduciary nuances of a multimillion-dollar supply chain is a challenging process that calls upon the instinct and intuition of the professional modeler. Not every analyst will generate the exact same mathematical model for the same set of demand, supply, and pricing data. The plans and insights generated by a comprehensive strategic network design study will inevitably be a function of both the quality of the software tool and the skill of the analyst who yields it. Therefore, we find it very useful to start with the center of gravity models to help modelers and decision makers later tackle the complexity of modeling a full supply chain with its costs.

In the following two problems, we will cover the physics center of gravity and the practical center of gravity. In both cases, we will be locating a single site.

Problem 1: Physics Weighted-Average Centering

Consider a hypothetical country based on the principles of rational organization. Let's call this country Logistica. The citizens of Logistica must choose a location for their capital.

Although all the citizens wish to live as near the seat of government as possible, it is impossible for all but a few of the citizens to relocate to their new capital. The map shown in Figure 2.1 shows the cities of Logistica, with the table showing the population. For more details on this problem, see the Excel file called `Logistica.xls` found on the book Web site.

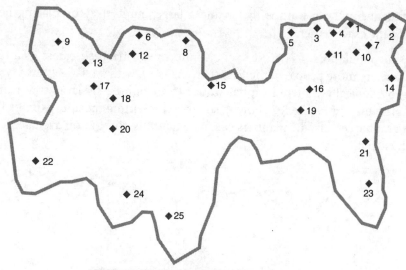

City #	Latitude	Longitude	Population
1	16.6	154.8	1,655,000
2	16.7	156.8	2,300,000
3	16.8	153.2	601,000
4	17	154	1,385,000
5	17	152	1,230,000
6	17.2	144.9	665,000
7	17.5	155.7	664,000
8	17.4	147.1	885,000
9	17.5	141.1	1,116,000
10	17.8	155.1	636,000
11	17.9	153.8	1,200,000
12	18	144.6	148,000
13	18.4	142.4	854,000
14	18.9	156.8	1,473,000
15	19.3	148.3	615,000
16	19.4	152.9	1,145,000
17	19.4	142.8	627,000
18	19.9	143.7	542,000
19	20.3	152.5	379,000
20	21.2	143.7	964,000
21	21.6	155.6	546,000
22	22.6	140.1	706,000
23	23.4	155.8	727,000
24	24	144.4	669,000
25	24.9	146.4	931,000

Figure 2.1   Map of Logistica and Population by City

The planners of Logistica first decide to place the capital in the most central location possible. Initially, they choose their capital location by considering the boundaries of Logistica, and then selecting the location that centers the country geographically. This would be analogous to a waitress balancing a large, empty tray by placing her hand underneath the exact center.

However, the population of Logistica is not uniformly distributed across its interior. In general, there are more people in the East than the West, with a relatively empty middle. Thus, a geographically centered capital would incur pointlessly large travel times for everyone. The map in Figure 2.2 shows a square representing the approximate location of the geographic center. The planners calculate that this point is, on average, 471 miles from each citizen.

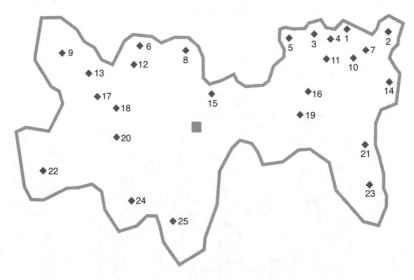

Figure 2.2 Approximate Geographic Center

Note: We can estimate the distance in miles between any two latitude and longitude points that are approximately halfway between the equator and the North or South Pole with the following equation:[1]

$$Dist(miles)_{ab} = 69\sqrt{(Long_a - Long_b)^2 + (Lat_a + Lat_b)^2}$$

$Dist(miles)_{ab}$ is the distance from *point a* to *point b*. $Long_a$, Lat_a, $Long_b$, and Lat_b are the longitudes and latitudes expressed as decimal numbers of point a and point b, respectively. If you want this measure to be in kilometers instead of miles, you simply change the 69 to 111.

---

1   *Designing and Managing the Supply Chain: Concepts, Strategies, and Case Studies.* David Simchi-Levi, Philip Kaminsky, and Edith Simchi-Levi. McGraw-Hill/Irwin. 2008. Page 87.

Ever rational, the Logistica planners decide to borrow an idea from their colleagues in physics and instead locate the capital based on a weighted average of locations of the cities already scattered about the country. Thus, a single large metropolis in the East would be given more consideration than a smattering of hamlets in the West. A city with one million Logisticans would have more sway than a city with just 500,000 Logisticans. This is how a physicist would calculate a center of gravity point.

Formally, this sort of problem can be solved with a simple closed-form equation. In practice, this means that the problem can be solved with a relatively simple calculation engine, such as an Excel spreadsheet, or even with pencil and paper. Mathematically, this also means that a computer program is guaranteed to solve this problem quickly.

Specifically, the mathematical formula for physics center of gravity (or the weighted average location) would find the coordinates of Logistica's capital as follows.

$$Lon_{cap} = \frac{\sum_{c \in C} P_c \, Lon_c}{\sum_{c \in C} P_c} \qquad Lat_{cap} = \frac{\sum_{c \in C} P_c \, Lat_c}{\sum_{c \in C} P_c}$$

Here, *Lon* represents a city's longitude, *Lat* represents its latitude, and *P* represents a city's population. Here we are using the population as the weighting factor. In other problems, many different weighting factors can be used. In network problems, customer demand is the most common. These formulas compute the weighted average of the longitude and latitude of all the existing cities (represented by the set C). The physics center of gravity for Logistica using the previous *Lon* and *Lat* equations is shown in Figure 2.3

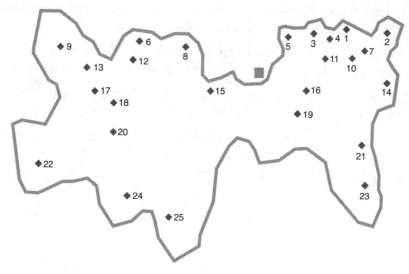

Figure 2.3   Logistica COG

With this approach, the weighted-average distance a citizen must travel is down to 388 miles. However, the location happens to fall in shark-infested waters offshore from a mountainous and deserted region. Obviously, the citizens of Logistica are not pleased with the resulting proposal. Though it's good for a physics calculation, the planners quickly realize that the center of gravity calculation would not prove practical.

Logistica has its share of uninhabited coastlines, mountainous regions, treacherous canyons, and forest preserves. These would not be good locations for the capital. The planners realize that they need to exclude these locations as options.

After further digging into this simple formula, the planners (and three of the four authors of this book) were surprised to find that the formula was not minimizing what they thought it was! The formula does not minimize the weighted average distance to each city. Instead, the formula minimizes the demand multiplied by the distance squared. That is, a city that is 100 miles away is weighted by a factor of 10,000 but a city that is 110 miles away (10% further away) is weighted by a factor of 12,100 (21% more). The principles of physics would naturally square distance when finding the center of gravity. However, to the citizens of Logistica (and to real world supply chain practitioners), this definition does not necessarily make sense. In general, if you have to travel 10% further, it should be about a 10% increase in cost. Because this is counter-intuitive, see the sidebar for an additional explanation.

In summary and after more discussion, the planners of Logistica came up with a list of reasons for why the physics center of gravity calculations should not be used:

- It does not minimize what you want to minimize: weighted average distance.

- It might point to the middle of a large lake, an ocean, the top of a tall mountain, a wildlife preserve, or the middle of a barren desert.

- It will almost never land in a location that exploits existing infrastructure, such as a city, a population center with a workforce, railways, highways, or ports.

- The method cannot take advantage of true road distances or travel restrictions—it has to rely on straight-line estimates based on latitude and longitude.

- It cannot be extended to include factors like costs, capacities, different types of facilities, different products, multiple levels of facilities (hub and spoke, suppliers, warehouses, retailers, etc.), or other practical considerations.

What the Physics Center of Gravity Actually Minimizes

The idea of computing a weighted-average position from a set of cities, and of treating this weighted average as some sort of unattainable "ideal point," has a deep-seated, intuitive appeal. There is a strong temptation to lend credence to the results of a clear mathematical equation. Indeed, when preparing and editing this book, the authors had to strongly resist describing the weighted-average technique as being some sort of hyperlogical, platonic ideal that is impaired only by a planner's inability to conjure up a "Brasilia/Canberra" city from scratch. There is something jarring about the realization that the weighted-average computation does not, in fact, minimize the total person-miles of our hypothetical country. With this in mind, we have included a brief digression on this particular aspect of center of gravity modeling.

Consider a slightly idealized version of the American state of Oregon. The bulk of the population of Oregon lives along the Willamette Valley, a fertile crescent that meanders for roughly 100 miles, from Eugene, Oregon, in the middle of the state, to Portland, Oregon, along its northern border. Let's simplify this situation slightly, and imagine that the entire population of Oregon lives in either Eugene or Portland, with a 4:9 ratio between the two populations and a total population of 1.3 million (we've made some assumptions to illustrate the point). Moreover, let's assume that the Willamette Valley (and the associated highway) runs 100 miles due north, from Eugene to Portland. Our COG problem is now nicely simplified—we only need to determine how far south from Portland to place our capital.

We can thus let X represent the distance from Portland to our capital. If $X = 0$ then Portland will be selected as our capital. If $X = 100$ then Eugene is the capital, and if $X = 50$ then the capital will be placed halfway in between the two cities.

Given our assumption that 400,000 people live in Eugene, and the remaining 900,000 live in Portland, the formula for computing total person-miles of travel will be as follows:

$$T = (9000000)X + (400000)(100-X)$$

Suppose we want to find the X that yields the smallest result for T, with the caveat that X must be between 0 and 100, inclusive. This result is easy to obtain from simple trial and error. If we let $X = 0$, then we obtain the result $T = 40,000,000$ (40 million). If we let $X = 100$, then we obtain the result $T = 90,000,000$ (90 million). If we let $X = 50$, then we obtain the result $T = 65,000,000$ (65 million).

In other words, if we place the capital in Portland, then, on average, an Oregonian will need to travel 31 miles to visit the capital (40 million person-miles divided by 1.3 million population). There is something jarring about the realization that the weighted average computation (the physics center of gravity calculation) does not, in fact, minimize the total person-mile (or with division by the total population, the weighted average distance a citizen must travel) of our hypothetical country. If we place the capital halfway between the two cities, we get a "split the difference" average trip of 50 miles.

Although these examples do not constitute a proof, it should be easy to see that we can minimize total person-miles simply by placing the capital in the larger city of Portland. Every location that is further away from Portland increases the total commuting distance for the citizens of Portland more than is offset by the total decrease in distance for the citizens of Eugene. This is true precisely because there are more Portlanders than Eugeneans.

However, consider the "physics weighted average" positioning of the capital. In our version of Oregon, we can simplify the map to a straight line (the same principles will apply on a globe using latitude and longitude).

With Portland at the 0 mile marker and Eugene at the 100 mile marker, the X variable again represents location of our capital. If we locate our capital using the physics weighted-average calculation, this value will be

$$X = ((900000)(0) + (400000)(100))/1300000 \approx 31$$

That is to say, a weighted-average position locates our capital about 31 miles south of Portland. The total person-miles for this location will be 55 million, or 43 miles per person.

This is clearly worse than the result obtained by placing the capital at Portland, because we are now requiring Oregonians to travel an average of 43 miles to the capital versus the 31 from before.

This shows that weighted average location computation (the physics center of gravity) does not minimize the average distance. It is minimizing something else!

The weighted average location computation is actually minimizing the demand times the distance squared. To show this, we will need a little calculus. Let S be the same as T except we are squaring the distance now. The formula becomes

$$S = (900000)X^2 + (400000)(100{-}X)^2$$

Now, as before, we might ask what value for X results in the smallest value for S. In this case, there are some relatively straightforward ideas from calculus that can be applied. Without getting bogged down in the mathematics, the minimization of S can be found by considering S to be a function of X, taking the derivative of S relative to X, setting this derivative to zero, and solving for X. This appears as follows:

$$S' = 2(900000)X - 2(400000)(100{-}X) = 0 \rightarrow$$

$$(2(900000) + 2(400000))X = 2(400000)(100) \rightarrow$$

$$X = 2(400000)(100)/(2(900000) + 2(400000)) \approx 31$$

Thus, when we place our capital at the weighted-average location that lies 31 miles south of Portland, we minimize not the total citizen miles (i.e., the T value) but rather the total citizen miles-squared (i.e., the S value).

For graphical evidence, see Figure 2.4. The x-axis represents the distance of the capital from Portland. So, 0 is at Portland, 10 would be 10 miles away from Portland, and 100 would be in Eugene. The Y-Axis on the left and the line with the square markers shows the demand weighted average distance. You can see that this steadily increases as we move the capital further from Portland. The Y-Axis on the right and the line with the diamond markers shows the total demand times distance squared (in millions). This is the physics center of gravity and you can see that its minimum point is about 31 miles from Portland.

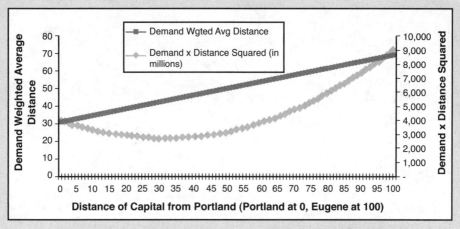

Figure 2.4   Graph with Different Center of Gravity Objectives

This sidebar has shown that the weighted-average position, while being most true to the "center of gravity" concept that is used by physicists, will minimize a value that does not have meaning for supply chain planners. Despite its intuitive appeal, it will give us an answer that is different from what we are actually trying to achieve. The physics center of gravity is good for balancing a tray or ensuring that flying airplanes are stable, but not very good for supply chain network design.

Problem 2: Practical Center of Gravity

The planners of Logistica, having learned that applying the physics center of gravity is flawed for network design problems, now try to find an approach to minimize the average distance traveled.

Some planners start to ask the question of what happens when they simply pick the three Eastern cities of 5, 16, and 11 (see Figure 2.5).

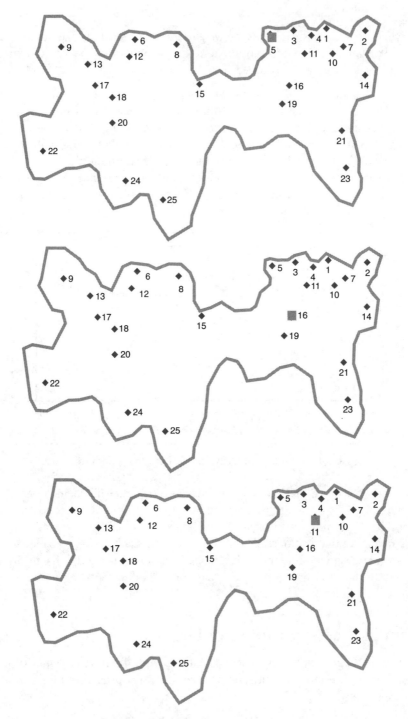

Figure 2.5   Picking Cities 5, 16, and 11

For each of these locations, a clever planner quickly calculates some additional key statistics to better understand these choices for a capital city. Figure 2.6 shows the weighted-average distance (miles) and the percentage of the population within 100, 200, and 300 miles of the city selected (shown in the column header).

	COG	City 5	City 16	City 11
Avg Distance	388	380	378	370
% within 100	0%	8%	7%	17%
% within 200	15%	32%	29%	39%
% within 300	33%	39%	48%	58%

Figure 2.6   Key Statistics for 5, 16, and 11

The planners of Logistica started to feel like they were on to a good approach. Maybe there is a lot more to this simple problem than they realized. First, they confirmed that the physics center of gravity solution (COG in the table) did not do very well on the key statistics for judging potential locations. It had a higher average distance and had very few citizens with 100, 200, or 300 miles radius. Because the objective of the physics center of gravity does not minimize average distance or percentage of customers close to the capital, this result is not surprising. The planners also realize that there may be trade-offs they have to make. For example, it may be important that a large portion of the population can easily get to the capital city, even if the average distance is a bit longer.

Of course, the curious planners want to know whether even better solutions existed. Because they are picking only one city as the capital, they can easily calculate the key statistics for all 25 cities. The full set of statistics can be seen in the spreadsheet, Logistica.xls. Figure 2.7 shows the statistics for a few key cities.

	COG	City 5	City 16	City 11	City 22	City 20	City 15	City 4	City 10
Avg Distance	388	380	378	370	800	585	428	381	399
% within 100	0%	8%	7%	17%	3%	7%	3%	24%	24%
% within 200	15%	32%	29%	39%	3%	12%	7%	48%	49%
% within 300	33%	39%	48%	58%	10%	23%	12%	56%	58%

Figure 2.7   Key Statistics for Other Key Cities

City 22 is at the extreme western point and, not surprisingly, does poorly on all the key statistics. Its citizens have grown used to the long travel distances to visit other Logistica cities and would not be surprised by this. Working our way east to the middle of the country, from City 22, we also analyze Cities 20 and 15. Neither does very well on the statistics, although 15 (which is close to the landmass center of the country) has a relatively decent weighted-average distance.

Cities 4 and 10 are much more interesting. Geographically, these cities are getting farther and farther away from any physics or mathematical center. However, they start to dominate on the percentage of the population that is within a certain distance. Both of these cities have 24% of the population within 100 miles—much higher than City 11's 17%. Likewise, City 10 dominates on the percentage within 200 and 300 miles. City 10 happens to be very close to the vast majority of the citizens of Logistica.

The planners quickly realize that comparing candidate locations allows them to exploit the richness of the problem (trading off average distance with percentage within a certain distance).

We now start to formulate a more realistic problem. We will formalize this in Chapter 3, "Locating Facilities Using a Distance-Based Approach." This problem selects the most central location from a list of possible candidates. Typically, the list of candidates is the entire list of cities. However, this is not a hard requirement, and a center of gravity problem could be forced to select from a subset of the cities, or could be allowed to select from among proposed locations that currently have no population at all.

You might think that by restricting the selection problem to a limited set of candidates, the problem would become easier to solve. In the case of just picking one city, this is true if you have the time to enumerate all the solutions like the planners of Logistica did. However, as you expand the problem, nothing could be further from the truth.

This more realistic and practical formulation of the problem is best solved with linear and integer programming techniques. This technique, which we will explore further in the next chapter, allows us to minimize the weighted-average distance (or other factors), picks locations we can actually use, and is very easily expanded to include other factors as well.

Lessons Learned from Center of Gravity Problems

In this chapter, we learned that locating a single point relative to demand is best done by minimizing the average weighted distance. You can accomplish this by picking your solution from a candidate list of locations. When picking one point, it is often simple enough to list out all the combinations and pick the best one. In the end, when you have a solution, it will tend to pull the single point close to as much demand as possible.

Although the physics concept of a weighted-average central point is appealing to our intuition and mathematically easy, it is flawed for network design studies. It can lead to undesirable locations as it is really minimizing the population multiplied by miles squared. Also, it cannot be extended to take advantage of existing infrastructure, road distance, or other facility costs.

End-of-Chapter Questions

1. Given the analysis presented in this chapter, where would you put the capital of Logistica and why? What factors went into your decision?

2. Besides the weighted-average distance and the percentage of customers within a certain distance of the capital, what other factors might the citizens of Logistica want to consider? Of these factors, which are quantifiable and which are qualitative?

3. If the planners of Logistica had been lucky enough for their first calculation to have picked an existing city, why should they still analyze other cities?

4. Name a reason why minimizing weighted-average distance is more important than maximizing the percentage of customers within a certain distance. Now, name a reason why maximizing the percentage of customers within a certain distance is more important than why minimizing weighted-average distance.

5. If instead of weighting the problem by the population of each city, assume that the analysis was done with each city having equal weight. That is, what matters is how close a city is to the capital, not how many people live there.

   a. Which capital location is now the best from an average-distance point of view?

   b. Which capital is now the best in terms of the number of cities within 100, 200, and 300 miles?

   c. Does this make the analysis easier or harder? Why?

6. What if the western part of Logistica decided it needed a capital city as well? Assume that City 15 was the easternmost city in this region. What is the best location for the capital of the western half of Logistica? Why?

7. Often, when logistics practitioners think about having some automatic way of picking a latitude and longitude, they think about a problem where the four cities form a square on a map. In this case, let's assume that each city has a population of one million and the latitude and longitude of the four points are (15,155), (15,158), (18,155), and (18,158).

   a. What is the latitude and longitude of the physics center of gravity?

   b. What is the average distance to each of the cities from this point?

   c. Is there a better point that would minimize the weighted-average distance?

   d. If the population of City 1 were five million, what would be the physics center of gravity latitude and longitude?

e. Sticking with the assumption in part (d), is there a location with a lower weighted-average distance? If so, where is the point?

f. Provide some reasons why the example in parts (a) to (c) (that is, four points, forming a square, each with equal demand) is not likely to occur in real-world supply chain network design problems?

8. In both the physics and the practical center of gravity, we used the straight-line distance from a city to the capital. Why could we use actual road distance in the practical center of gravity but not in the physics center of gravity?

3

LOCATING FACILITIES USING A DISTANCE-BASED APPROACH

As you saw from the examples in the introduction, supply chain network modeling can handle very sophisticated problems. To be able to model, optimize, and develop good solutions to complex supply chains, you need to understand the basics.

In this chapter, we will build on the problem structure developed in Chapter 2, "Intuition Building with Center of Gravity Models," for Logistica. We will adapt and expand on what we learned by now applying it to a more realistically sized example from the retail industry. We will start with the most basic supply chain network model: optimally selecting "P" facility location(s) considering only distance. This will be your building block for further analysis throughout the rest of the book. You will find that this first step is quite valuable, and some real-world projects may not require more detail.

Retail Case Study: Al's Athletics

Let's now begin our analysis of a major retailer in the United States specializing in sporting equipment and apparel named Al's Athletics. In the late 1960s, Al Alford, a young football coach in a small Texas town many miles away from a major city, decided that the kids in his town shouldn't be held back because they didn't have access to the right sporting equipment to practice with. Al felt every kid in his town deserved a chance to become the next great Texas-born sports star. And opening a store close to them to supply the right equipment could be the first step for these kids to grow up to be the next Nolan Ryan, Mean Joe Green, or Olympic gymnast Kim Zmeskal. Before Al knew it, his business was booming. He began selling to parents not only for their children but also for themselves. Al and his family quickly realized that they could repeat their success within similar cities across the south, and 50 years later Al's Athletics has grown to be one of the largest retail sporting-goods chains in the U.S. Al's Athletics now has major retail-store outlets in 41 U.S. states.

Al's grandson (Al the third) is now the CEO and lives and works near the original store in Brownfield, Texas. The family began using a warehouse in nearby Lubbock, Texas, about ten years ago, and the Brownfield Store and many others in Texas have realized great savings in transportation costs and provided significantly better service ever since. Al (the third) realizes that competition in the sporting-goods industry is high, and he has always believed in the philosophy of his grandfather that "proximity to the customer means everything in this business." As a result, Al wants to look at adding and optimizing warehouse locations across the country to lower costs and provide the best service of any sporting goods chain in the U.S.

In Chapter 2 we used the practical center of gravity analysis to determine a best location for Logistica's capital city. In this chapter we will build on this idea by now selecting optimal "P" warehouses from a predetermined list of options.

In practice, as mentioned in Chapter 2, most supply chain network design modeling will begin with a set of "potential" facility locations from which to determine the optimal. Let's begin helping Al by reviewing the formal problem definition for this approach.

Formal Problem Definition for Locating "P" Facilities

Given a set of customer locations and their demand, find the best *P* number of facilities (plants or warehouses) that minimize the total weighted distance from the facility to the customer, assuming that each facility can satisfy the full demand of the customer and that all demand is always satisfied.

Let's begin our understanding of this definition by starting with the customers, their location, and their demand.

When modeling a supply chain, the *customer* refers to the final delivery location for our products. We are only concerned with the final delivery point that is relevant for the supply chain we are analyzing. The product may travel farther before being consumed, but if it is beyond the scope and control of the supply chain we are analyzing, we are not concerned with it.

Depending on the type of supply chain we are modeling, the customer can refer to many types of locations. Each situation is different, and you may need to use various definitions of customers in your supply chain model. Here is just a sample of what a customer location may represent:

- If we are modeling a retail supply chain, the customers may be each of the store's locations. (The supply chain of Al's Athletics follows this logic.)

- If the retailer sells directly to their online customers and delivers to the home, the customers may be each ZIP or postal code they ship to. (Later, we will talk about aggregation strategies and the use of postal codes in place of actual home addresses.)

- If the retailer has stores in a shopping mall, they may often ship to the store in a truck that is shared with other retailers in the same mall. To do this, all the retailers ship to a common location, often referred to as a pool point, and then the truck goes from there to the shopping malls for a set fee. In this case, because the assignment of stores to this truck will not change, you may model the pool point as the customer location.

- If we are modeling a manufacturing firm that sells to retailers or wholesalers, the customers could be the warehouses of the retailer or wholesaler. As mentioned previously, from the point of view of the manufacturing firm, they only need to get the product to these warehouses—this is the end of the supply chain for them. The retailers or wholesalers will take it from there.

- If the manufacturing firm ships to other manufacturing firms (say, a steel company shipping to an automaker), the customers will be the other manufacturing sites.

- If a firm's supply chain already has a fixed and unchanging set of warehouses with a fixed set of customers they service (and this will not be changed), we can model these warehouses as the customer points.

- If a firm's supply chain services the home-building industry, the customer may represent the job site where we need to provide the product.

- If a company's supply chain exports product and we are not responsible after the product leaves the country, the customer can be the port of exit.

When we define and discuss customers, we are considering each of the unique ship-to locations as a customer point. For example, a manufacturing company that sells product to a large retailer needs to consider each of the large retailer's locations they ship to, not a single geographic point representing the retailer, such as the address of the headquarters or a single billing address commonly found within financial data.

"A picture speaks a thousand words" could not be more applicable than to the review of any company's demand map or customer network. Visualizing customers on a map is a powerful way to start to analyze any supply chain network. In fact, companies can get a lot of value just from this exercise. Let's start by looking at the customer network for Al's Athletics.

Figure 3.1 shows Al's customer points plotted on a map of the U.S. This simple analysis allows you to quickly see where your customers are located, the geographic span of your customer base, and the number of areas you are delivering to.

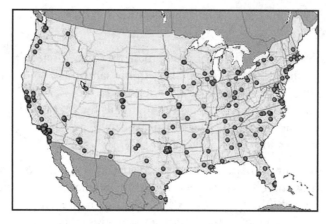

Figure 3.1   Al's Athletics Customer Map

We can quickly see that Al's has stores in most states and a large concentration of stores in Texas (and especially the Dallas area), California, and some points along the East Coast.

To visualize Al's customers on a map, we needed to link every customer point to latitude and longitude values. This process is called geocoding. Some firms have this data readily available internally. Otherwise, network modeling software can do this for you or there are many Web sites that can provide this data as well.

Now, we need to add demand to the model.

As discussed in Chapter 2, regarding the selection of the capital city of Logistica, optimal location selection often does not make sense based on geography of destination locations alone. The citizens recognized that they wanted to weigh the importance of all city locations by their relative population. Supply chain design studies work very similarly to the way the citizens of Logistica selected their capital; however, these studies use historical or forecasted amounts of product to be delivered to each of the customers as opposed to population. We call this the "demand." Initially, our models will use demand to determine the relative importance of each customer. For example, a customer who receives 300 truckloads per year of product will be more important than one who receives only 50 truckloads. Of course, in later chapters we will explore the weighting of the importance of customers by other measures as well. But, for now, this assumption will serve us well because it is generally less expensive to be closer to your 300 truckload customer than to be closer to the one with 50.

To specify each customer's demand, we need to determine the appropriate unit of measure. When thinking in terms of financial systems, we are concerned with the total revenue associated with each customer. However, this unit of measure is not sufficient for network design. In this case, we need a measure that will help us correctly apply the associated manufacturing and logistics costs and capacities. For example, many companies refer to transportation costs in terms of "cost per truckload" and production capacity in terms of "units per hour." If demand is measured in terms of revenue, we have no way to determine how much "revenue" fits in one truckload or how many dollars of revenue may be produced in one hour. Determining the best unit of measure for demand, production, and transportation is an important step at the beginning of all supply chain models.

For now, let's assume that we will measure Al's demand at his stores in terms of total weight (pounds). For example, Store #1 may have a demand of 750,000 pounds of product per year. Al's selection of weight is probably the most common unit of measure in network design due to the fact that most transportation costs are based on the weight. In later chapters we will discuss modeling demand with multiple products and with different units of measure for transportation, warehousing, and manufacturing cost. Besides weight, other common units of measure include

- Total pallets
- Total cube (physical space the product occupies)
- Total truckloads
- Total cases

This list is just a sample. As you build more complex models, you may find you need to use different units of measure.

The first question Al will face when beginning his demand data collection will most likely be, "What time horizon should we analyze?" Typical network design studies use an entire year's worth of demand. In addition to being a common reporting time period for many companies, reviewing an entire year of activity within a supply chain ensures that all peaks and valleys of your customers' buying patterns are accounted for in your analysis.

Imagine a retail company that bases its network design decisions on demand from only a three-month period (January through March). The holiday season has ended and shoppers are hibernating until spring arrives. For some retailers it may be common for their total purchases during this period to be significantly lower than those during the peak holiday months. Designing their network on this data might result in a recommendation to decrease their number of warehouse locations in order to save on capital costs. This company will most likely enjoy their cost savings until the fourth quarter, when a major spike in holiday sales leaves them far short of capacity and facing many problems including lost sales. The cost of this misstep will most likely far outweigh all previous savings. This company will quickly realize that a valuable study needs to incorporate both peak *and* off-peak periods of demand.

Based on this discussion, Al has provided you with a year's worth of demand data for the 200 cities in the U.S. where Al's Athletics stores are located. On the map in Figure 3.2,

Al's customers are sized by their relative demand, and in the second map, we see demand displayed by each state's relative shading (the deeper the shading, the more demand within that state).

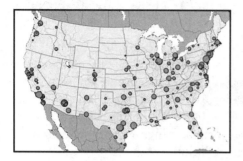 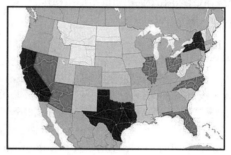

Figure 3.2   Al's Athletics Customer Maps Formatted by Relative Demand

Returning to our formal problem, if we were asked to write out the current problem that Al is trying to solve, it would be stated as follows:

"Find the best *P* number of facilities (warehouses) that minimize the total weighted distance from the facility to the customer, assuming that each facility can satisfy the full demand of the customer and that all demand is satisfied."

With this definition, you can see we are going to assume that our measure of "best" is that we are minimizing the weighted-distance to customers.

We are also going to assume that we are picking our facilities from a predefined set of locations. We will provide Al with a list of the top 25 most frequently selected locations for warehouses in the U.S. to use as a potential list to select from. Adding their current location in Lubbock, Texas, as another potential option, their choices for facilities are shown in Figure 3.3.

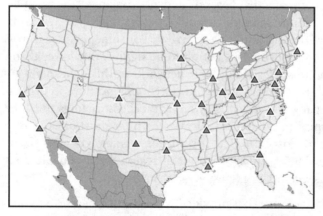

Figure 3.3   Map of All Potential Warehouse Locations for Al's Athletics

With a predefined set of facilities, we can then build a matrix with the distance between each facility and each customer. In most cases, you are interested in the road distance between two points. This can be easily and automatically calculated with commercial-grade network modeling software.

Formulating and Solving the Problem

We now need a method to solve Al's problem and others like it.

In Chapter 2, we saw how finding the single-best point was relatively easy. We simply listed all the possible combinations and picked the best one. This would not be too difficult for Al's problem either, because it would result only in 26 choices.

However, we should not be fooled. The problem gets more complicated very quickly. When we pick the best two sites, we need to figure out the best two locations and determine which customer is served by each of these two sites. Right now, it is easy to assign the customers to the two locations we have chosen—simply pick which of the two is the closest one. However, the complication lies in the number of potential combinations we must consider to prove this. If we have 25 potential facility locations and need to find the best 5, there are 53,130 combinations to evaluate. That is, you would have to evaluate every possible combination of 5 different facilities. The formula for this in Excel is `=COMBIN(# of potential facilities, # you want to pick)`.

Figure 3.4 shows that the number of combinations explodes as you consider more potential facilities. If you add up the total number of combinations for picking between the best two to ten facilities, the total comes to more than seven million options. In this case, there are too many choices to simply enumerate every combination with a spreadsheet.

Total Potential Facilities	25
Number Picked	**Total Combinations**
2	300
3	2,300
4	12,650
5	53,130
6	177,100
7	480,700
8	1,081,575
9	2,042,975
10	3,268,760

Figure 3.4   Evaluating 25 Potential Facilities Versus Resultant Combinations

The number of combinations can quickly get out of hand if we consider more choices as well. If Al's Athletics wanted to evaluate 10 times more facilities, the number of combinations they would need to consider for all combinations with two to ten facilities picked would grow by more than 32 billon times the original (see Figure 3.5). In other words, a 10-times increase in the potential facilities leads to a 32-billion-times increase in the number of combinations that need to be considered. It doesn't take much to find yourself with a potential network design problem containing many more combinations of facilities than the number of atoms in the universe.

Total Potential Facilities	250
Number Picked	Total Combinations
2	31,125
3	2,573,000
4	158,882,750
5	7,817,031,300
6	319,195,444,750
7	11,126,241,217,000
8	337,959,576,966,375
9	9,087,357,513,984,750
10	219,005,316,087,033,000
11	4,778,297,805,535,250,000

Figure 3.5   Evaluating 250 Potential Facilities Versus Resultant Combinations

Based on this information, you can easily see the need for the use of linear and integer programming techniques for sorting through these combinations in a systematic fashion to find the best answer. See the sidebar, "Permutations and Combinations," for more information.

Permutations and Combinations

A casual reader, when confronted with a statement like "There are 53,130 ways to pick the best 5 facility locations from a set of 25 candidates" will tend to take the author at his or her word. There are some diligent readers who will dutifully verify that their Excel spreadsheet displays COMBIN(25, 5) = 53,130 when programmed correctly. Precious few will actually recall the formula that COMBIN applies, and double-check that it is in fact appropriate for this instance.

That's a shame, because the underlying mathematics here is actually quite intuitive. Understanding the difference between permutations and combinations is useful, in as much as it helps us understand why optimization problems can be so difficult to solve. Moreover, a reader who readily recalls such mathematics has a head start on quickly assessing the difficulty of a wide variety of real-world, combinatorial problems.

To begin with, let's consider a similar, related problem to the one at hand. Suppose we are trying to select players for a "line" in ice hockey. For those uninitiated into the joys of "Canada's game," an ice hockey line consists of three offensemen who skate together. The center, left wing, and right wing all know each other's styles and habits, and will enter and leave the ice as a trio.

Suppose there are five aspiring players who want to play on the first line. The coach must not only choose which players will skate in the first line, but which positions they will play. Mathematically, this is an example of what's known as a k-permutation, or a k-permutation of n. The terminology is actually pretty straightforward—*k* refers to the number of items to select, *n* refers to the size of the pool from which the selections are made, and the word "permutation" is used because the selections are made into positions.

(A true mathematician would perhaps be more likely to say that one uses permutations when "order matters," but the difference between "order" and "distinct positions" is a matter of semantics. We were careful to choose hockey as our example because the right wing, the left wing, and the center are considered three distinct positions. When you are selected to skate the right wing, you'd best not skate out to the left side of the rink.)

At any rate, let's give our five aspiring players useful labels. We will refer to these players as White, Orange, Blonde, Pink, and Blue. Our coach (perhaps himself Mr. Brown) has five choices for who is to play left wing. Suppose he selects Blonde to play this position. The coach now has four choices for who is to play center—one fewer than before, because Blonde cannot play both positions simultaneously. We will assume that the coach selects Blue to center the line. He now has three choices to play right wing (because Blonde and Blue are both occupied). We'll let this spot go to Orange.

We can write out this selection as (Blonde, Blue, Orange). There are, in fact, 60 such selections for the coach to choose from. The number 60 can be deduced from the number of choices the coach faced at each step: $5 \times 4 \times 3 = 60$.

Although it might seem tedious, let's write out all 60 choices. We will write them out in a very particular way, in order to serve a purpose that will soon become apparent. For purposes of space, we'll also shorten their names somewhat.

(Blnd, Blue, Orng)	(Blue, Blnd, Orng)	(Blnd, Orng, Blue)	(Blue, Orng, Blnd)	(Orng, Blnd, Blue)	(Orng, Blue, Blnd)
(Blnd, Blue, Whte)	(Blue, Blnd, Whte)	(Blnd, Whte, Blue)	(Blue, Whte, Blnd)	(Whte, Blnd, Blue)	(Whte, Blue, Blnd)
(Blnd, Blue, Pink)	(Blue, Blnd, Pink)	(Blnd, Pink, Blue)	(Blue, Pink, Blnd)	(Pink, Blnd, Blue)	(Pink, Blue, Blnd)
(Blnd, Orng, Whte)	(Blnd, Whte, Orng)	(Orng, Blnd, Whte)	(Orng, Whte, Blnd)	(Whte, Blnd, Orng)	(Whte, Orng, Blnd)
(Blnd, Orng, Pink)	(Blnd, Pink, Orng)	(Orng, Blnd, Pink)	(Orng, Pink, Blnd)	(Pink, Blnd, Orng)	(Pink, Orng, Blnd)
(Blnd, Pink, Whte)	(Blnd, Whte, Pink)	(Pink, Blnd, Whte)	(Pink, Whte, Blnd)	(Whte, Blnd, Pink)	(Whte, Pink, Blnd)
(Blue, Orng, Whte)	(Blue, Whte, Orng)	(Orng, Blue, Whte)	(Orng, Whte, Blue)	(Whte, Blue, Orng)	(Whte, Orng, Blue)
(Blue, Orng, Pink)	(Blue, Pink, Orng)	(Orng, Blue, Pink)	(Orng, Pink, Blue)	(Pink, Blue, Orng)	(Pink, Orng, Blue)
(Blue, Pink, Whte)	(Blue, Whte, Pink)	(Pink, Blue, Whte)	(Pink, Whte, Blue)	(Whte, Blue, Pink)	(Whte, Pink, Blue)
(Orng, Pink, Whte)	(Orng, Whte, Pink)	(Pink, Orng, Whte)	(Pink, Whte, Orng)	(Whte, Orng, Pink)	(Whte, Pink, Orng)

Again, these are the 60 choices an ice hockey coach has for selecting the first line of his hockey team, when selecting from among a pool of five players.

Now, suppose that rather than selecting the offensive line for hockey, our coach is selecting the three Chasers for a Quidditch team. (If you are reading this book in the year 2050, you might need to be reminded that Quidditch was the game played in the Harry Potter series. This game was later adapted for "Muggle play" by people with lots of creative energy and too much free time—i.e., college students.) The Chasers play a similar offensive role in Quidditch as do the offensemen in ice hockey. However, there is no distinction between the different Chasers. That is to say, a Chaser would not refer to himself as being a "left Chaser" or "right Chaser," but simply as playing a Chaser.

Thus, when selecting Chasers, the coach would not assign his choices to specific positions. Rather, he would simply name the three players he has chosen. This is an example of an unordered selection, as opposed to selecting hockey players for an offensive line, which is an ordered selection.

Mathematically, the Quidditch-Chaser problem is an example of what's known as a k-combination, or a k-combination of n. Again, the terminology is straightforward. k refers to the number of items to select, n refers to the size of the pool from which the selections are made, and the word "combination" is used because the selections are made into an interchangeable squad. (Or, more formally, because order doesn't matter.)

In this case, if the coach were to select the same first-rate athletes as before, then we would write them out as {Blonde, Blue, Orange}. Here, we use {} instead of () because an unordered grouping of players is simply a set, whereas an ordered grouping of players is a "tuple." (Although the word *tuple* seems too silly to be real, it is in fact how computer scientists refer to an ordered set. Each one of our 60 offensive lines is a distinct tuple of three players.)

The formula for computing the number of choices there are in a k-combination problem is based on the more straightforward formula we used for the k-permutation problem. Look at the clever way we wrote out our 60 selections above. We arranged them in a grid, with 6 columns and 10 rows. Each row contains the same three players, shuffled into six different combinations. The fact that there are six different combinations per row can be demonstrated from a simpler k-combination problem—if you are restricted to merely selecting the postions for your line from a pool of only three players, then you select from among three choices for left wing, followed by a selection for two choices for center, with right wing being determined by the process of elimination.

Thus, to determine the number of choices for a k-combination of n, we first compute the number of choices there are for a k-permutation of n, and then divide by the number of choices for the k-permutation of k. That is to say, to determine the number of ways to select Chasers from a five-player pool, we first compute the number of ways to select an offensive line from a five-player pool, and then divide by the number of ways to select an offensive line from a three-player pool. The answer in this case is $\frac{5 \cdot 4 \cdot 3}{3 \cdot 2 \cdot 1} = 10$. It is probably worthwhile to write out these 10 subsets of our five-player pool.

{Blnd, Blue, Orng}

{Blnd, Blue, Whte}

{Blnd, Blue, Pink}

{Blnd, Orng, Whte}

{Blnd, Orng, Pink}

{Blnd, Pink, Whte}

{Blue, Orng, Whte}

{Blue, Orng, Pink}

{Blue, Pink, Whte}

{Orng, Pink, Whte}

To formulate these equations in a more general way, we just need to consider the way we counted our coaches' options for selecting the offensive line. We can note that

$$5 \cdot 4 \cdot 3 = \frac{5 \cdot 4 \cdot 3 \cdot 2 \cdot 1}{2 \cdot 1} = \frac{5!}{2!}$$

where the ! refers to the factorial formula (i.e., $5! = 5 \times 4 \times 3 \times 2 \times 1$). Thus, the general formula for k permutations of n is $n \cdot n-1 \cdot n-2 \cdot \ldots \cdot n-k+1 = \frac{n!}{(n-k)!}$

We can then exploit our observation that k combinations of n is just k permutations of n divided by k permutations of k. Thus, we can derive the Excel formula that started this sidebar:

$$COMBIN(n,k) = \frac{\frac{n!}{(n-k)!}}{\frac{k!}{(k-k)!}} = \frac{n!}{k!(n-k)!}$$

Note that this last step took advantage of the equation *(k-k)!=0!=1*. The idea that zero factorial equals one is essentially accepted by definition, and is useful for other aspects of mathematics besides the derviation we demonstrated here.

Finally, knowing the formula for COMBIN (and even better, seeing this formula derived) gives one a gut instinct as to why selection problems can so easily become intractable. At the heart of the COMBIN function lies the factorial function, and the factorial function has a "look and feel" that is very similar to a function that is known for its (sometimes lethal) rapid growth, the exponential function. Indeed a classic mathematical formula, whose proof we will spare you, is based on bounding the factorial formula. Although this formula has multiple names and variations, we will present it as a simplification of Stirling's formula:

"Simplification of Stirling's Formula"

$$n! \geq 2.5 \cdot \sqrt{n} \cdot \left(\frac{n}{3}\right)^n$$

From this formula, we can see that the factorial function is capable of growing very, very quickly, because it will outgrow the complex function to the right, which embeds the exponential function (to see this, test it out in Excel). This is sufficient cause for anyone mathematically savvy to be wary of enumerating the choice selections that are implied by the COMBIN function. That is to say, it is not by accident that we chose to discuss hockey lines and Quidditch Chasers, because more complicated examples would have easily outstripped our ability to type, much less ponder.

Let's begin by systemically analyzing Al's Athletics' problem, based on the criteria we outlined in Chapter 1, "The Value of Supply Chain Network Design." This will be a generic analysis that can be applied to any similar problem.

Let's start with the key *data* elements and define some terminology:

- We have a set of customers we need to serve. We will index this set by the uppercase letter J. So in the case of Al's Athletics, the set J would be {New York, Chicago, Los Angeles, and so on}. We will call out an individual customer with the lowercase letter j.

- We know the demand of each customer. We will call the demand d and indicate the demand for customer j as d_j.

- We have a set of potential facilities we can select from. We will index this set by the uppercase letter I. An individual facility will be designated by the lowercase i. In the case of Al's Athletics, the facilities are the 25 warehouses.

- We can represent the distance matrix to show the distance from facility i to customer j as $dist_{i, j}$.

The *objective* of this problem (and Al's objective) is as follows:

> Minimize the average weighted distance from the warehouses to the stores. This is equivalent to minimizing the total weighted distance. However, it is more intuitive to report on the average weighted distance.

We also have a few *constraints* or rules that we must keep in mind:

1. We have to meet all the demand. This one seems obvious, but we do not want the model to outsmart us by minimizing the distance to a customer by not serving the customer at all.

2. We are limiting the number of facilities to P. For example, Al's wants to know the best two, three, four, and so on. If we did not limit the number, the model would likely select all facilities. The more facilities, the closer Al's may be to the customers (our overall objective).

We have two *decisions* we are making:

1. Do we use the warehouse at location i? We will denote this decision as X_i. If $X_i = 1$, then we elected to use the facility at location i; if $X_i = 0$, then we elected to not use this facility. This is what is called a binary variable. It can take on a value of only 0 or 1.

2. Does facility i service customer j? We will denote this decision as $Y_{i,j}$. If $Y_{i,j} = 1$, then facility i will service customer j. If $Y_{i,j} = 0$, then facility i will not service customer j.

We can now formulate this as a mathematical program. Don't be intimidated by the math here. If this is new to you, this is just a way to systemically describe the problem. You won't need to understand the mechanics, but instead use the math equations to help build intuition. When we formulate it like this, a standard mixed-integer programming engine can determine the solution. Solving these problems is the subject of many other books and courses and we will not cover the techniques here, although we will provide an example in Excel later in this chapter. However, to understand how these models work, where they become difficult, and how you can implement the results in practice, it is important that you develop good intuition of the underlying concepts.

One way to think about this mathematical formulation is that it gives us a set of rules to make sure we solve the problem correctly. For example, the constraints force us to select the right number of facilities and to meet all the demand. The objective function then guides the search to find a solution that meets our definition of what is best.

The first thing we want to formulate is the objective function. This function gives us the ability to determine whether one solution is better than another. In this case, we want to minimize the total weighted distance from warehouses to customers. So the objective function is to *minimize* the following equation (1):

$$\sum_{i\in I}\sum_{j\in J} dist_{i,j}\, d_j Y_{i,j}$$

The sigma sign (Σ) means that we sum whatever comes next (which is why Excel uses this symbol for their calculation button). The letters under the Σ tell us exactly what we are summing up. The "$\in$" symbol stands for "element of." In the first summation, we have $i\in I$. So this means that we sum over all the individual facilities in our complete set of potential facilities. The second summation has $j\in J$. This tells us to sum over each of the individual customers in our entire customer set. So the entire formula tells us to sum over all facilities and all customers to determine the single answer.

Note that only $Y_{i,j}$ is a variable. This means that we need to try different values for each of the $Y_{i,j}$ variables. The other terms are input values.

The attentive reader will quickly see why we need constraints. If we simply set every $Y_{i,j}$ to zero (0), the result from the equation will be zero. Zero is certainly a small number However, mathematical optimization techniques are clever and will quickly realize that if we set the $Y_{i,j}$ variables to negative numbers further and further from zero, we will continue to minimize the result of the function. So we see that we need to prevent both of these cases to ensure that the mathematical optimization returns a solution that we can actually use. We will prevent this by now including our constraints.

The first constraint will stipulate that every customer must be fully served. Here, we have an equation for every customer. By using the $\forall$ symbol, which means "for every," we can write all these equations simply as (2):

$$\sum_{i\in I} Y_{i,j} = 1; \forall j \in J$$

So, starting with the end of the equation, $\forall j \in J$ simply means for every customer in the overall set of customers. The first part of the equation states that we sum up all the values of $Y_{i,j}$ for every potential facility and that sum must equal 1. Think of the 1 as representing 100%. So this says that 100% of every customer's demand must be served. This prevents the problem we mentioned in the objective section where we simply set the $Y_{i,j}$ variables to zero and didn't serve demand. This now forces the problem to serve all demand.

The next constraint stipulates that we want to locate exactly P facilities. The constraint is (3):

$$\sum_{i\in I} X_i = P$$

The existence of this equation might beg the question: Shouldn't the optimization engine actually tell us the optimal number of facilities? This is a good question, but for this problem we are interested only in minimizing the average distance to service a customer. So, we have not and do not want to apply a cost of adding a warehouse in the objective function. Because warehouses in this formulation are "free" and we are minimizing just the total weighted distance, there would be nothing preventing the optimization from selecting every warehouse. Is this a flaw in the model? No. We do not want to complicate our model at this stage and the constraint limiting the number of warehouses solves the problem nicely. Besides, our analysis will actually allow us to review a lot of extra information on the value of adding each additional warehouse to our network. This analysis shows that you do not need to directly include every important factor in your mathematical model—through clever analysis, you can often get good information on an important factor without actually modeling it. (You will see this in later chapters as well.)

We still need a way to say that if a warehouse serves a customer in constraint (2), then that warehouse must be considered "open" or "selected."

For this, we need a way to tie together constraints (2) and (3). If we didn't have this constraint, the mathematical optimization would simply make good assignments for the $Y_{i,j}$ variables and then randomly pick P warehouses to open. That is, the model may serve the New York stores from the New York warehouse, but decide not to open the New York Warehouse. We display the constraints that make this link as follows (4):

$$Y_{i,j} \leq X_i; \forall i \in I, \forall j \in J$$

Note that we have a constraint for every customer and every facility. If you had 200 customers and 25 warehouses, you would have 5,000 different constraints. (This standard notation allows us to compactly write this as one line.)

We read this constraint (4) as saying that we can't make an assignment of a warehouse to a customer unless we open that warehouse. That is, if the X_i variable is zero, we cannot assign customers to that particular warehouse.

Now, we need to force the values of X and Y to make sense. In this case we will force these variables to take on a value of either 0 (zero) or 1 (one). For the X variables, this simply means that we either open a warehouse at location i ($X_i = 1$) or we do not ($X_i = 0$). It does not make sense to open half of a warehouse. For the $Y_{i,j}$ variables, in this case, we are setting their values to either 0 or 1 as well. In this case, it could, in fact, be logically fine to service a fraction of customer j's demand from a warehouse. That is, a customer could receive product from multiple warehouses. For now, however, we will set the constraint to require single sourcing of each customer from only one warehouse. Later, we will analyze the impact of this decision and how you can change it. Here are constraints (5) and (6):

$$Y_{i,j} \in \{0,1\}; \forall i \in I, \forall j \in J$$

$$X_i \in \{0,1\}; \forall i \in I$$

When we incorporate all the previously described objectives, decisions, and constraints, here is the mathematical problem formulation of the complete model:

Minimize
$$\sum_{i \in I} \sum_{j \in J} dist_{i,j} d_j Y_{i,j}$$

Subject to:

$$\sum_{i \in I} Y_{i,j} = 1; \forall j \in J$$

$$\sum_{i \in I} X_i = P$$

$$Y_{i,j} \leq X_i; \forall i \in I, \forall j \in J$$

$$Y_{i,j} \in \{0,1\}; \forall i \in I, \forall j \in J$$

$$X_i \in \{0,1\}; \forall i \in I$$

One nice thing about a formulation written like this is that you can solve this for any given set of data. Although we wrote it with Al's Athletics in mind, it would be equally applicable to other companies. If you solve this with a quality commercially available network design tool, this mathematical formulation will be taken care of for you inside the application. But it is important for you to understand what the network design tool is doing so that you can develop better models and have a clear understanding of the solutions being provided.

It is also important to note what is hard about this problem.

As we saw at the start of this section, there are a lot of different combinations of potential warehouses. There are simply too many combinations to enumerate all the possibilities as we did when we were evaluating the single best location for Logistica. What makes the number of combinations difficult is that there are no reliably efficient techniques for finding the optimal solution when you have binary variables that take on either a zero or a one.

In most cases when you have only discrete variables, the problem falls into a class of problems known as NP-Hard. Without getting into the details and for the purposes of this book, this means that there are no known algorithms that are guaranteed to solve the problems to absolute optimality in a reasonable time as the model size grows. Because this is an engineering and business book, we will take the optimization and computing power as a given and develop ways to build our models so that we get good answers that we can implement.

Certainly, contemporary optimization technology and computing problems allow us to solve very realistically sized problems. We will discuss the size of "reasonable" problems and modeling approaches to ensure that they are solvable throughout the remainder of the book. And based on what we learned previously, we will also keep in mind that the more we avoid adding discrete variables to our model when possible, the better off we'll be and the larger the models we will be able to solve.

Hands-On Excel Exercise

Another way to think about the formulation of the model is to see it formulated in Excel. The entire formulation with referenced row and column headings can be seen in the Excel file called `MIP for 9-City Example.XLS` found on the book Web site. The formulation in Excel will not match exactly with our previous algebraic formulation, but it will give us valuable additional insight.

This is a simple model with very few data points so that it can be solved with Excel's built-in solver. We picked nine cities as customers and used these same nine cities as our potential warehouses.

You can see that the objective function is cell B11, as shown in Figure 3.6. This is the sum of cells B2 to B10. Each cell within B2 to B10 is the sum of demand times the SUMPRODUCT of the appropriate $Y_{i,j}$ column (for Chicago it would be B28 to B36; for Boston it would be J28 to J36) with the matching column in the distance column (for Chicago it would be B41 to B49; for Boston it would be J41 to J49). This is one way we capture the double summation in Excel.

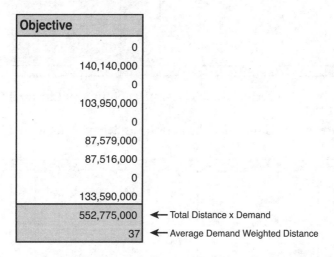

Figure 3.6　Objective Function Portion of MIP for Nine-City Excel Model

The cells C15 to C23 correspond to our X_i decision variables, as shown in Figure 3.7. If we want to place one of the warehouses in the solution, we fill in the corresponding cell value with a 1. We put in a 0 if we do not choose this site.

Set of Cities	Demand-i	X-i (use = 1)
Chicago	2,870,000	1
Atlanta	572,000	0
New York	8,450,000	1
St. Louis	350,000	0
Detroit	901,000	1
Cincinnati	333,000	0
Pittsburgh	306,000	0
Charlotte	723,000	1
Boston	610,000	0
Total Facilities		4
Total Allowed		4	← Constraint on number of facilities allowed

Figure 3.7 X_i Decision Variables Portion of MIP for 9-City Excel Model

The $Y_{i,j}$ variables are B28 to J36, as shown in Figure 3.8. The customers are along the top row and the warehouses are listed in the A column. So if we want the St. Louis customer to be serviced from Chicago, we would fill in a 1 in cell E28. All the other cells in the column for St. Louis should be zero.

Ship [from,to]	Chicago	Atlanta	New York	St. Louis	Detroit	Cincinnati	Pittsburgh	Charlotte	Boston
Chicago	1	0	0	1	0	0	0	0	0
Atlanta	0	0	0	0	0	0	0	0	0
New York	0	0	1	0	0	0	0	0	1
St. Louis	0	0	0	0	0	0	0	0	0
Detroit	0	0	0	0	1	1	1	0	0
Cincinnati	0	0	0	0	0	0	0	0	0
Pittsburgh	0	0	0	0	0	0	0	0	0
Charlotte	0	1	0	0	0	0	0	1	0
Boston	0	0	0	0	0	0	0	0	0
Must Be Supplied	1	1	1	1	1	1	1	1	1

Figure 3.8 $Y_{i,j}$ Decision Variables Portion of MIP for 9-City Excel Model

And the distance matrix is shown at the bottom, as depicted in Figure 3.9.

Distance Matrix	Chicago	Atlanta	New York	St. Louis	Detroit	Cincinnati	Pittsburgh	Charlotte	Boston
Chicago	0	720	790	297	283	296	461	769	996
Atlanta	720	0	884	555	722	461	685	245	1,099
New York	790	884	0	976	614	667	371	645	219
St. Louis	297	555	976	0	531	359	602	715	1,217
Detroit	283	722	614	531	0	263	286	629	721
Cincinnati	296	461	667	359	263	0	288	479	907
Pittsburgh	461	685	371	602	286	288	0	448	589
Charlotte	769	245	645	715	629	479	448	0	867
Boston	996	1,099	219	1,217	721	907	589	867	0

Figure 3.9   Distance Matrix Portion of MIP for 9-City Excel Model

Let's now take a closer look at the solver (see Figure 3.10). Open the solver dialog box (this requires the use of Microsoft Excel's "Solver" add-in, see the book Web site for further instructions on how to install and access this functionality), here you will see how we tell Excel's Solver about the problem. At the top, we fill in B11 and click the "Min" radio button in the "Set Target Cell" and "Equal To" sections. This tells the Solver to minimize cell B11.

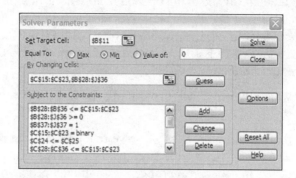

Figure 3.10   Excel Solver for MIP for 9-City Excel Model

The section of the dialog box called "By Changing Cells" is where we tell the solver what our decision variables are. This is what the solver can change.

Next, we fill out our constraints. Unfortunately, the solver does not allow you to order the constraints. So within the list you see that we have constraints that the cells B28 to J36 (the $Y_{i,j}$ decision variables) are set to $>= 0$. We do not want these to be negative. In this particular problem, these variables will naturally be either a 0 or 1, so we do not have to specify that these will be binary. (Note that this is a special case for this problem and not a general rule.) This is constraint (5).

The cells C15 to C23 (X_i decision variables) are set to binary. So they will be either 0 or 1. This is constraint (6).

We also have a constraint for B37 to J37 to equal 1. This says that each of these cells must be 1. This tells us that we have to meet 100% of the demand. This is constraint (2) from earlier.

We also have a constraint for (3) stipulating that C24 is less than or equal to C25. This tells the model how many facilities it can pick.

As you may recall, constraints (4) within the mathemical formulation are the most complicated and this remains true when creating them in the solver as well. These constraints state that we cannot assign a customer to a warehouse if that warehouse is not open.

To do this in Excel, look at the table that corresponds to the $Y_{i,j}$ variable. Each column represents a customer and each row represents a warehouse. So if we place a "1" in a particular cell, the warehouse corresponding to the row of that cell must be open (otherwise we would have a logical problem of a customer receiving product from a warehouse that was not open). Also note that we've created a single column to indicate whether a warehouse is open or closed (this is the X_i table in cells C15:C23). Now, to set the constraint, we just specify that the values in X_i table must be greater than or equal to the values in any single column in our $Y_{i,j}$ table. That is, if we want to put a "1" somewhere in the $Y_{i,j}$ table, we had also better put a "1" in same row of the X_i table. This ensures that we simultaneously make the open and close decisions (in the X_i table) and the assignment decisions (in the $Y_{i,j}$ table). We need to add this constraint once for each column in our $Y_{i,j}$ table—that is, we need to put in this constraint for every customer. These are written as B28:B36 <= C15:C23, C28:C36 <= C15:C23, D28:D36 <= C15:C23, and so on up to J28:J36 <= C15:C23.

This model can now be solved. We will explore this model more in the end-of-chapter questions.

As you will see in the next section, after we solve this problem the resulting solution, when mapped, displays a "star-burst collection" pattern that is characteristic of strategic network design solutions in general (and of center of gravity problems in particular).

Analysis of This Model for Al's Athletics

Let's now continue our analysis of this problem for Al's Athletics.

Our goal is to determine the best P warehouses but we are not assuming that we know the best value for P. Instead, we will run multiple scenarios with different values for P and compare the answers.

In this analysis, we will run the best 1, 2, 3, 4, 5…10 warehouses for Al's Athletics supply chain. You can see the resultant networks in Figures 3.11 through 3.15.

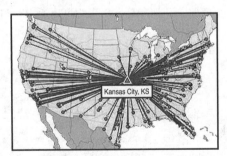 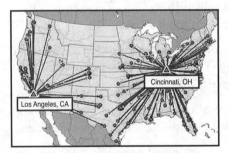

Figure 3.11   Al's Athletics' Best 1 and Best 2 Warehouse Solution Maps

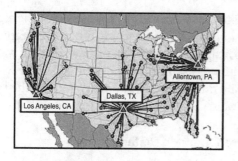
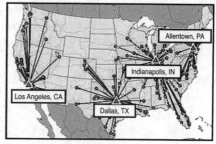

Figure 3.12 Al's Athletics' Best 3 and Best 4 Warehouse Solution Maps

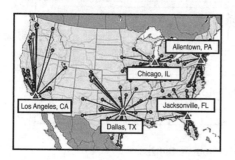
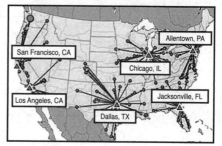

Figure 3.13 Al's Athletics' Best 5 and Best 6 Warehouse Solution Maps

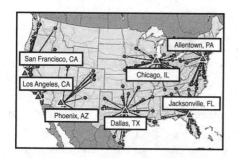
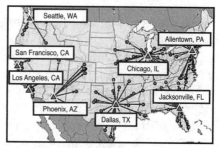

Figure 3.14 Al's Athletics' Best 7 and Best 8 Warehouse Solution Maps

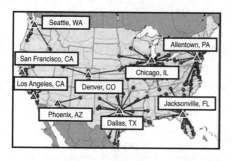
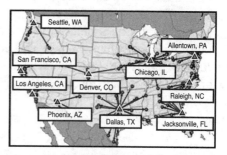

Figure 3.15 Al's Athletics' Best 9 and Best 10 Warehouse Solution Maps

We can plot the objective function (in weighted-average distance) for each of the solutions. As you can see from the graph shown in Figure 3.16, there are diminishing returns to adding more facilities. This gives us some insight into the value of an additional warehouse. When we go from three to four facilities, distance decreases by 31%, but when going from four to five, distance decreases by only 19%. We still do not know the best answer, but we now have more insight into value of adding an additional warehouse.

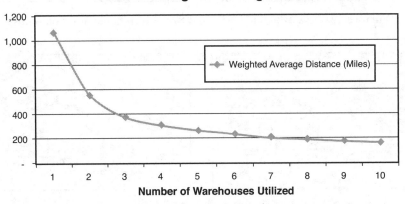

Figure 3.16 Weighted Average Distance Comparison Across Solutions

There is other information we receive from comparing Al's solutions as well. Weighted average distance is a good summary of service level. However, we can learn more by examining the amount of demand served within certain distance bands or ranges. Figure 3.17 uses bands of 100, 400, 800, and 3,200 miles. In Al's case, the 100 miles is used to represent demand that is very close to the warehouse, the 400 miles is used for demand reached within a single day, 800 miles for two days drive, and 3,200 miles for all other demand. To read the table in Figure 3.17, look at the 1 Warehouse Solution row. You can see that 1% of the demand is within 100 miles, 4% of the demand is within 100 to 400 miles, 30% of the demand is within 400 to 800 miles, and the remaining 65% is over 800 miles away. In reviewing the full table, we see that solutions improve within each distance band. Demand served within 100 miles, for instance, increases with each new optimal warehouse added to the network. It is important to note that these measures do not necessarily have to improve because they are not part of the objective. But, these distance bands are closely correlated with improving the average distance. That is, by improving the average distance, you are most likely going to have more customers closer to warehouses.

Number of Warehouses	Distance Band (Miles)				Weighted Average Distance
	100	400	800	3200	
1 Warehouse	1%	4%	30%	65%	1,072 Miles
2 Warehouses	12%	23%	39%	26%	558 Miles
3 Warehouses	18%	38%	35%	9%	383 Miles
4 Warehouses	31%	38%	24%	7%	311 Miles
5 Warehouses	36%	38%	23%	3%	266 Miles
6 Warehouses	41%	36%	22%	1%	236 Miles
7 Warehouses	45%	37%	17%	1%	210 Miles
8 Warehouses	47%	38%	15%	0%	193 Miles
9 Warehouses	49%	38%	13%	0%	178 Miles
10 Warehouses	51%	40%	9%	0%	165 Miles

Figure 3.17   Distance Band Comparison Across Solutions

Al now has the information he needs to make a decision. Of course, Al realizes that being closer to his stores and therefore the customers will provide better service and help reduce his transportation costs from the warehouses to the stores (sometimes called "outbound" transportation). He also knows that adding warehouses comes with the cost of extra facilities, extra management time, extra administrative overhead, and extra complexity. (He will have more inventory to manage and suppliers will have to ship to more locations.)

However, Al gets a lot of insight into these costs and complexities with this analysis. He knows that going from one warehouse to two warehouses significantly improves his ability to quickly get product into his stores. And although he might not be able to quantify costs, he knows that going from five locations to six only gets him from 97% to 99% of demand within 800 miles. He may not think this small increase in demand within 800 miles is worth the extra cost of an additional warehouse.

Al still does not have an easy decision. He may still want to do some additional financial analysis on the alternatives before making a final selection. But, for many retailers, having warehouses as close to the stores as possible is the single most important objective. Often, the only question they must answer is exactly how many warehouses and where to locate them.

Lessons Learned from Locating Facilities with a Distance-Based Approach

When you optimize by allowing multiple sites and your objective is to minimize the weighted-average distance, the optimization locates the facilities close to demand. That is, large demand points and areas with a lot of demand will tend to pull locations closer.

This model seems to have a lot of assumptions and a restricted objective function. It is only minimizing the weighted distance. However, do not sell this model short.

First, and most importantly, it is a model that can solve many supply chain network modeling problems. For some firms, ensuring that you are as close to your customers and demand as possible is the most important consideration.

Second, distance is a good measure of your ability to quickly deliver. The closer your facilities are to your customers, the faster you can get product to your customers (assuming your facility has the product ready to ship). So in a supply chain in which response time to customers is important, the model in this chapter may be all that is needed. This is often true for many retailers and similar businesses. Often, the decision making in an organization will require additional financial analysis, but you may be better off doing this outside of your optimization runs.

Third, using these types of models help us realize exactly how much value is gained by each additional facility in your supply chain. These scenarios allowed us to demonstrate to Al that each additional warehouse location offers less benefit to the network on the whole. Knowing exactly how much the benefit diminishes with each additional location allows Al to understand the business implications of adding more warehouses and offering better service to his customers. This helps him "optimize" the Alford family adage of "Service Is Everything."

Fourth, distance and transportation costs are highly correlated (the farther you have to drive, the more expensive it is). So this model is a good approximation for minimizing the transportation costs to the customers. And, in many cases, this is the most expensive transportation leg.

Finally, although there is a desire for the optimization run to give us a single correct answer, the world is more complicated than that. In practice, when you're making an important strategic decision, it is more important to run multiple scenarios to understand the trade-offs, and understand the marginal value of adding facilities. This information can then be coupled with strategic business objectives to make a final decision.

Ultimately, this simple model provides a lot of value and can be used as building block for more complicated models.

End-of-Chapter Questions

1. Al's Athletics has just started the data collection required for you to help them develop their associated network design model. During the first conference call, they indicate that they have three reports they feel could be used to determine the model's demand per customer. Based on what you have learned in this chapter, which of the following reports could best help us create the demand file? For each of these reports, describe why or why not you could use it to create the demand file.

   a. Financial Report listing invoice payments received from all customer accounts over the course of the past year.

   b. Warehouse Management System Reports showing inventory levels during each month of the first quarter of last year.

   c. Transportation Shipment Report showing origins/destinations, shipment weight, date, and cost over the course of the past year.

2. Al's data provided us with demand in terms of pounds. Why does this make a good unit of measure for our model? What other units of measure may have been acceptable and why?

3. In the mathematical formulation, what does the X decision variable mean? Why does it have to take on a value of 0 or 1?

4. In the mathematical formulation, why can Y be a continuous linear variable and still give a legitimate solution? (For example, what does a Y value of 0.45 mean?)

5. If you have a problem with 1,000 customers and want to consider 1,000 potential warehouse locations, how many different combinations of choices will you have if you want to pick the best 10 warehouse locations, or the best 25 warehouse locations?

6. Complete the following questions using the MIP for 9-City Example.xls spreadsheet found on the book Web site and reviewed previously in this chapter:

   a. What are the best two locations with the given demand?

   b. What are the best three locations with the given demand? How much did the average distance decrease?

   c. What are the best three locations if you consider the demand for each of the cities to equal 100,000? How does this solution compare to part (b)?

   d. What are the best three locations if the demand for every city is equal to 100,000 except New York and Chicago, which are 200,000 each?

7. "Selecting the Best Two to Four Warehouse Locations for Al's Athletics—Mini Case Study"

This hands-on example will give you your first opportunity to further your understanding of the book's concepts by building and running scenarios within a commercially available software package. You can get details on the software and data files from the book Web site. The data file you need is called Al's End of Chapter Questions.zip. This file contains additional information to help you with the case study. You will be asked to create and review a slightly altered version of Al's Athletics supply chain network, as well as prepare and run scenarios in order to determine the network's best two, three, and four warehouse locations.

a. What are the locations of the best two, three, and four warehouses selected for Al's network?

b. For each scenario, what are the average throughputs per warehouse? Which warehouse processes the most demand? Which warehouse processes the least?

c. Prepare a quick chart depicting how much demand is serviced within 100-mile, 400-mile, and 800-mile distance bands within each solution. How much more demand can we service within 100 miles utilizing the best four warehouses as opposed to the best two?

4

ALTERNATIVE SERVICE LEVELS AND SENSITIVITY ANALYSIS

In this chapter we want to expand on the models from Chapter 3, "Locating Facilities Using a Distance-Based Approach," by adding some sophistication to modeling service levels. We will also dig deeper into the important concept of sensitivity analysis and learn why running many different iterations of your model will ultimately bring the most value to your project. Finally, this chapter will introduce the notion of infeasible models and efficiently finding their root causes.

What Does Service Level Mean?

Picture yourself standing in the kitchen after a long week of work or classes. You are starving and rifling through your delivery menus from a few of your favorite pizza parlors. Pat's Pizzeria has always had your favorite deep dish, but there is only one location all the way across town and it always takes at least 60 minutes for your order to arrive. You then come across the most recent menu left at your doorstep for a new place called Primo Pies. You have heard good things about them and know they have three locations in town, with one right down the street. You quickly pick up the phone and place your order for a deep-dish pizza from Primo Pies. Within 30 minutes you are on the couch enjoying your pizza and watching your favorite movie. Pure bliss!

The rationale you just used to determine which restaurant to order pizza from is very similar to the decisions consumers make in almost all product markets. Think back to how Al's Athletics got their start in the industry. Locating themselves in a market where the other big sporting-goods stores couldn't provide an acceptable service level led to their success in the industry. For all retailers, this means locating stores close to customers or locating warehouses that ship online orders close enough to get next-day service (or be willing to pay for air freight).

The reasons customers want stores close is similar to why businesses want to be able to service these stores from nearby locations. Al valued being close to customers in two senses. He wanted his stores in convenient locations for his customers, but he also wanted his warehouses close to his stores. Shoppers at Al's may not care how far away the warehouses are. But Al knows that the closer his warehouses are, the faster he can replenish the stock at stores and ensure that shoppers have the product that they want when they want it.

Supply Chain Design Service Levels

The term "service level" can mean many different things in supply chain modeling. If we think back to the pizza example, we can relate the most important measures for network design.

Let's assume that the pizza parlors in town understand that many of their customers go through the same decision-making process you went through. If they are thinking of where to open their first or a new restaurant, they will likely consider one of the following two definitions:

1. **Minimize the average distance**—We have seen this definition before. This is the same one used within our previous practical center of gravity solutions. In this case, for a given number of restaurants, you want to minimize the distance and thus increase the ability to get pizza to as many hungry people as fast as possible so that they don't go elsewhere.

2. **Maximize the percentage of customers within a certain distance**—There is a rationale that says customers don't care if the pizza gets there in 12 minutes or 18 minutes. That is, demand won't change within a range of times. But if you cross some threshold, maybe 30 minutes, demand will drop off dramatically. So, when locating facilities, we care more about being within 30 minutes of the most potentially hungry people rather than minimizing the overall distance to the customers.

These are the two definitions used to measure service level for companies completing strategic network design studies across all industries and geographies. For example, if a chemical company or consumer products company wants to be within one day of their customers, they will both need to consider their facility locations in terms of one or both of the above service-level definitions.

There are two other measures of service level that are important to the business, but they are not directly relevant to network design. We are discussing them here to help you understand why these are not inputs to an accurate network design model:

1. **Fill Rate**—This is a measure of the percentage of orders that are filled from inventory. In the case of our pizza places, the fill rate would measure the percentage of time a customer calls up looking for a sausage pizza but they don't

have any sausage on hand. This will obviously hurt sales and it is clearly important for a firm to maximize the fill rate. However, the location of the pizza parlor has nothing to do with whether they have enough of all the key ingredients when customers want them.

2. **Late Orders**—This is a measure of how late the shipment is. To measure this factor, we look at when a customer orders a product, when the firm promises a delivery, and whether that delivery actually happens on time. In the case of our pizza parlor, if a customer places an order and they are promised delivery within 20 minutes (we originally assume it will take ten minutes to cook), but their pizza doesn't show up until 40 minutes later (because in actuality it took us 30 minutes to cook), this results in a late order that is not the fault of location of the pizza parlor.

In summary, the best way to think about network design is that it gives you the opportunity to meet your service promises. If you want to open up pizza parlors with a 30-minute delivery promise, network design is the best way to put your parlors in the right locations. If you can't keep enough sausage on hand or if you take 30 minutes to cook a pizza, you can't blame the location of the facility. Other more tactical processes, like training of employees, defining reorder policies, or managing supplier service levels, will have to be altered to improve these measures in the long run.

Consumer Products Case Study: Chen's Cosmetics

Let's now begin service-level analysis of a major consumer products company in China. Chen's Cosmetics specializes in supplying affordable quality cosmetics to stores across China.

Chun Chen grew up in Beijing near one of the most popular fashion-modeling agencies in China. She envied the gorgeous models leaving the studio each day looking nearly perfect thanks to the cosmetics their makeup artists had so artfully applied before they had their pictures taken. Chun soon became a makeup artist herself and found that the best cosmetics were quite costly and could never be afforded by most people. So she decided to work together with her brother, a talented chemist, to produce a line of professional quality cosmetics for affordable prices to be sold all over China.

After years of trials, formulations, and production out of a single manufacturing facility, Chun Chen, now CEO of Chen's Cosmetics, finds herself with more demand than the existing plant can produce. Chun decides she will add two production facilities in locations across China and wants to ensure that these additional locations offer the best service to Chen's existing customer base (distributor's warehouse locations across China).

Chun and her supply chain team first started their analysis by reviewing the service level they currently offered from their single production facility in Guangzhou, China. Before they could do this, however, they had to formally define how they would measure service level.

Service levels within network design are most commonly measured by transit time or distance. Gathering data on transit times for all existing and potential lanes is not always an easy task for companies, however. This data can be quite variable, and requests from carriers to provide detailed data in this area are often denied or unreliable. Network design analysts commonly use a distance equivalent to transit time to avoid the frustrations with transit-time data. As we continue to review samples of these concepts, however, bear in mind that any assumptions we will make regarding speeds of vehicles and driver service hours are all estimates for the purposes of understanding the concept and are not based on definitive research on our part or any public standards and guidelines.

If we think about our previous example determining where to locate pizza restaurants within a 30-minute delivery range, it's safe to assume that a modeler would not look to determine the time it takes to get to each and every house within the city. Instead, she makes an assumption about how fast a delivery vehicle will travel on average (including required stops for traffic signals, congestion, and so on) and how much time it takes to make the pizza (say, 15 minutes in this case), and then determines the max mileage that can be covered with that speed in 15 minutes (the remaining time to get to the house).

For instance, if a delivery vehicle can travel at 20 mph on average, we would assume that our objective should be centered around demand within 5 miles:

20 mph * .25 hour (15 minutes for delivery) = 5 miles traveled

Chun Chen and her team made similar assumptions in their model when deciding on additional plant locations. Due to the tight operating hours of the distributors' warehouses requiring trucks to travel during peak periods of city traffic, in conjunction with the requirement for trucks to travel on a significant number of underdeveloped roads as they approach distributor warehouses, the team knew that the number of kilometers that could be covered within a day would be limited. Using averages from their own shipment data as well as public information about average speeds and traffic patterns on both major roadways and less-traveled local access roads, the team determined that delivery trucks would travel at an average speed of only 50kmph. Combining this information with an assumption that drivers can only work a maximum of eight hours per day, we are able to calculate our equivalent one-day transit distance:

50kmph * 8 hours ~ 400 kilometers traveled in one day

Therefore, Chun's team may now easily extrapolate this to their desired two-day service level as a distance band of 800km from plants outbound to distributor warehouse locations. Modeling their "As Is" state (or baseline) shows Chun that she is currently able to service only 18% of her demand within two days of service (800km).

In addition, on average Chen's Cosmetics products travel a whopping 1,603km before reaching the customers (see Figure 4.1). Therefore, Chun realizes that these new manufacturing locations not only will alleviate their capacity problems but also should have a significant impact on their ability to offer more competitive service levels.

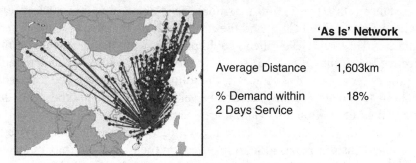

'As Is' Network

Average Distance   1,603km

% Demand within   18%
2 Days Service

Figure 4.1   Baseline Network for Chen's Cosmetics

The supply chain team has previously determined that the cost to open a manufacturing facility in any of its 24 potential locations is fairly equal. Therefore, they will simply focus their modeling on maximizing the service they can provide to their customers. The existing network with current Guangzhou plant (the dark rectangle with a flag on top in the far southeast) and all potential plant locations (the lighter rectangles with flags on top) can be seen layered on top of the customer base (the circles) within Figure 4.2.

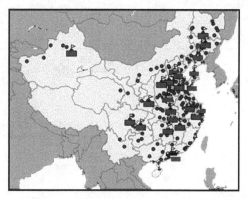

Dark Plant = Existing Guangzhou Location
Light Plants = Potential Plant Locations

Figure 4.2   Potential New Plant Locations for Chen's Cosmetics

The supply chain team decides to run models using each of the two popular network design service-level objectives discussed previously and will then analyze the results with the company's executives. This strategy represents the beauty of network design modeling. You don't have to pick just one potential objective or set of data. You can easily try several different optimization runs (also known as scenarios) and then compare the results. Sometimes, you will be pleasantly surprised by what you find.

The two scenario results, shown in Figures 4.3 and 4.4, use the objective of minimizing average distance (Objective 1) and maximizing the demand within 800km (Objective 2). But, the analysis showed Chun and the executive team that there is little difference in the resultant 2-day service level Chen's could provide by implementing the results from one objective versus the other. That is, one solution covers 80% of the demand in two days and the other covers 78%. However, the solution that is optimized for average distance is about 10% better on this metric (588 versus 654). Chun and her team decide to add their two additional plants (Tianjin and Nanjing) found with Objective 1 (minimizing the average distance). They have proven that offering the best service to their highest-demand customers has little to no effect on their ability to maximize service to markets within 800km, and therefore these locations are ideal for their expansion.

Objective 1: Offer the best service possible to the customers who purchase the most

Objective 2: Maximize the demand you can service within 800kms/2 Day Service

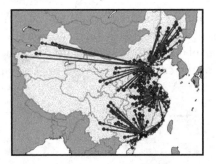 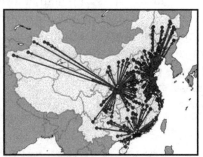

Figure 4.3   Solution Map Comparison for Chen's Cosmetics Service-Level Scenarios

	"As Is" Network	Objective 1	Objective 2
Average Distance	1,603km	588km	654km
% Demand within 2 Days Service	18%	78%	80%

Figure 4.4   Scenario Results Comparison Between the Two Objectives

Based on this study, it would have been easy for Chun to assume that there was no need for the alternative service-level analysis based on the very similar results produced by each objective. But let's now look forward five years. The new plant locations were a huge success and Chen's Cosmetics' market share grew so much that the company now has interest from markets in 38 European countries as well. Chen's Cosmetics has enough production capacity to handle this growth but obviously cannot afford to fly smaller amounts of product directly from their plant to each of the approximately 500 customer distribution centers in Europe that have already promised their business. The question for the company now becomes "Where should they locate three regional warehouses across Europe?" These will store and deconsolidate Chen's Cosmetics products flown or shipped in from the China plants before distribution to all European distributor locations.

Consumer Products Case Study: Chen's Cosmetics European Warehouse Selections

Chen's has elected to hire a third-party logistics (3PL) company to handle all the transportation of product inbound and outbound from their regional European distribution centers. Chen's team has worked with the 3PL and jointly identified 48 warehouse location options from which they are to select the optimal 3 they would like to use.

Chen's team must now conduct a modeling exercise very similar to the one performed five years ago to locate their additional plants in China. Their initial network map, depicted in Figure 4.5, shows their 500 customer locations plotted with their 48 potential warehouse locations.

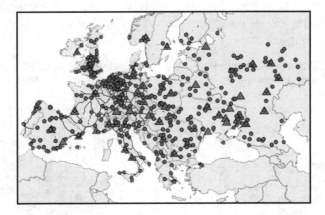

Figure 4.5   Chen's European Market

The team initially creates a model with the objective of maximizing service level by minimizing the distance weighted by demand, as had been so successful in their previous study in China. In this case, however, information provided from the 3PL tells them that their delivery vehicles are able to travel faster than the estimate that Chen's team made for their deliveries in China. The service that the 3PL offers Chen's Cosmetics includes

---

their ability to cover 500km in one day of transit. Therefore, maximizing the two-day service level within this model is now equivalent to a distance of 1,000km.

As seen in the results shown in Figure 4.6, this study selects locations in Paris, Rome, and Kremencug (in the Ukraine). These locations will offer only 65% of their new markets serviced within two days (1,000km). This doesn't seem like an ideal solution to the team, because the company is just getting their feet wet in these completely new market areas. Chun quickly reminds them, however, that there are other methods of optimizing their service level—and that although they were committing to only three warehouses continent-wide, they wanted to make sure they could offer good service to as many markets as possible. This can only help them grow sales even faster during their first few years in the market. She wants to make sure that they test their European service strategy fully before making a decision.

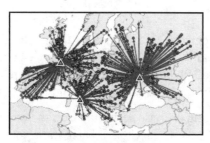

Objective 1

Average Distance	709km
% Demand within 2 Days Service	65%

Figure 4.6   Chen's European Market "Objective 1" Scenario Results

The team then shifts their model strategy slightly to look at the selection of locations based on maximizing the demand they can service within two days, and produces the results shown in Figures 4.7 and 4.8.

Objective 1: Offer the best service possible to the customers who purchase the most

Objective 2: Maximize the demand you can service within 1,000kms/2 Day Service

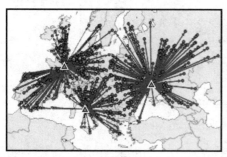

Selects:
Paris, France
Rome, Italy
Kremencug, Ukraine

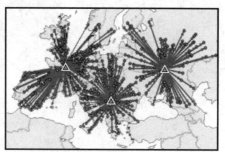

Selects:
Paris, France
Belgrade, Serbia
Voronezh, Russia

Figure 4.7   Solution Map Comparison for European Market Service-Level Scenarios

	Objective 1	Objective 2
Average Distance	709km	729km
% Demand within 2 Days Service	65%	84%

Figure 4.8   Scenario Results Comparison Between the Two Objectives

The team is astonished to find that they can increase their service within two days from 65% to 84% of their demand with this new recommended structure of warehouses in Paris, Belgrade, and Voronezh. Just by quickly adjusting and running another version of their model, they are able to find a solution that allows them to serve an additional 19% of customer demand at a competitive service level. Chun pitches this solution to the executive team as not only optimal for the current customer base but also as a way to put them in a better position to grow their business with these clients in the near future. Chen's is now ready to make their entrance into the European cosmetics market!

Mathematical Formulation

Let's return to our mathematical problem formulation we first defined in Chapter 3. Now that we have introduced a new model objective (maximizing demand within a distance band), let's see how we adjust this formulation to get us to our new modeling goal. The constraints remain the same. So we will just focus on the new objective function. To maximize the amount of demand with a certain distance (or time), we can write this:

$$\text{Maximize} \quad \sum_{i \in I} \sum_{j \in J} (dist_{i,j} > HighServiceDist\,?\,0:1) d_j Y_{i,j}$$

In this new equation, the term $dist_{i,j} > HighServiceDist$? simply is a test condition that asks whether the distance between the two points is greater than the service parameter. In the previous case, 800km was the service parameter in China. The next part of the expression simply tells the model what to do. If the statement is true, this expression takes on a value of 0. If it is false, it takes on a value of 1. So points that are more than 800km get a 0 and do not help our objective function. This then guides the objective to look for as many combinations as possible in which the demand is within the 800km.

In the next section, we will also introduce a constraint that forces the average customer demand to be within a certain distance (or time) of the servicing facility. This constraint tells the solver engine that even if a customer cannot be assigned to a facility within the *HighServiceDist* restriction, you would still prefer that it be assigned to a facility that is reasonably close. Note that in the previous formulation, we are providing no guidance for how to assign customers outside the *HighServiceDist,* so, it will make a random selection.

$$\sum_{i \in I} \sum_{j \in J} dist_{i,j} d_j Y_{i,j} < AvgServiceDist \sum_{j \in J} d_j$$

The *AvgServiceDist* constant represents the largest average customer demand assignment distance that will be tolerated. If this constraint is tight for a given solution, you know that the typical unit of demand travels exactly this distance along the final leg of its journey. It would be reasonable to choose an *AvgServiceDist* value that is reasonably close to your *HighServiceDist* optimization target. That is, you might try to maximize the amount of customer demand that is serviced within 800km, while insisting that the average servicing distance is no larger than 1,000km.

If we provide the full formulation, it would look like this:

Maximize $$\sum_{i \in I} \sum_{j \in J} (dist_{i,j} > HighServiceDist ? 0 : 1) d_j Y_{i,j}$$

Subject To:

$$\sum_{i \in I} \sum_{j \in J} dist_{i,j} d_j Y_{i,j} < AvgServiceDist \sum_{j \in J} d_j$$

$$\sum_{i \in I} Y_{i,j} = 1; \forall j \in J$$

$$\sum_{i \in I} X_i = P$$

$$Y_{i,j} \leq X_i; \forall i \in I, \forall j \in J$$

$$Y_{i,j} \in \{0,1\}; \forall i \in I, \forall j \in J$$

$$X_i \in \{0,1\}; \forall i \in I$$

Note that the *AvgServiceDist* constraint, while discouraging the solver from selecting lengthy demand assignments, doesn't strictly prohibit such selections. Our MIP engine might still assign a few stray demand points to unreasonably distant servicing facilities, while nevertheless maintaining a reasonable average servicing distance across the supply chain as a whole. Indeed, for demand points that are fairly small, these "outlier" selections will tend to crop up with some frequency. Because small demand points don't contribute much to the left-hand side of our *AvgServiceDist* constraint, the solver has little incentive to provide them with reasonable service.

To address this problem, it is common to apply constraints that limit the maximum servicing distance. Whereas the average service distance can be modeled with a single constant applied to a single constraint, the maximum service distance constraint requires a single constant applied over many constraints. Luckily, these constraints are all very similar, and thus can be written out concisely in mathematical notation:

$$Y_{i,j} \leq (dist_{i,j} > MaximumDist\,?\,0:1); \forall i \in I, \forall j \in J$$

This family of constraints simply says that customer j cannot be assigned to facility i if the distance from i to j is larger than the *MaximumDist* constant value. As you might expect, we need to be more relaxed with our selection of *MaximumDist* than we are with our selection of *AvgServiceDist*. (Indeed, if *MaximumDist* is smaller than *AvgServiceDist*, then the *AvgServiceDist* constraint will be redundant. Can you see why?) A reasonable selection of values for our three constants might be *HighServiceDist* = 800km, *AvgServiceDist* = 1,000km, and *MaximumDist* = 1,200km.

In addition, a sophisticated modeler might want to minimize the average servicing distance, while treating the quantity of demand met with high service as a constraint. For example, suppose that the user determines, by solving the model we've been working with, that it is possible to assign 75% of demand to a facility within 800km, while maintaining an average servicing distance that is no greater than 1,000km. By flipping the constraint and the objective, he or she can instead minimize the average servicing distance, while insisting that 75% of demand be met with a high-service shipping lane.

Suppose the result of this second model is a solution that has an average servicing distance of 950km, while also meeting 75% of demand with a servicing facility within 800km. This solution will have an interesting property, in as much as it will have successfully optimized two different goals. That is to say, it might be possible for a solution to meet more than 75% of demand with high service, but only by sacrificing the average servicing distance and allowing it to go higher than 950km (indeed, higher than 1,000km). Moreover, it might also be possible for a solution to achieve an average servicing distance lower than 950km, but only by allowing more than 25% of demand to be met through a low-service distance.

If a solution has this multiple-goal property, under which one cannot improve one objective without degrading the other, it can be called Pareto optimal (named after an Italian economist). One can always stumble into a Pareto optimal solution by using this "two solve" process, which involves using the objective result of one solve as the constraint on the second. However, there are more general, and more powerful, techniques for surveying the entire range of Pareto optimal choices. We shall discuss these methods in more detail in Chapter 11, "Multi-Objective Optimization."

Here is an overview of the two complementary models that can be used to create a solution that is Pareto optimal with respect to average servicing distance and total demand met with high service.

First Model

$$\text{Maximize} \qquad \sum_{i \in I} \sum_{j \in J} (dist_{i,j} > HighServiceDist ? 0 : 1) d_j Y_{i,j}$$

Subject to:

$$\sum_{i \in I} \sum_{j \in J} dist_{i,j} d_j Y_{i,j} < AvgServiceDist \sum_{j \in J} d_j$$

$$Y_{i,j} \leq (dist_{i,j} > MaximumDist ? 0 : 1); \forall i \in I, \forall j \in J$$

$$\sum_{i \in I} Y_{i,j} = 1; \forall j \in J$$

$$\sum_{i \in I} X_i = P$$

$$Y_{i,j} \leq X_i; \forall i \in I, \forall j \in J$$

$$Y_{i,j} \in \{0,1\}; \forall i \in I, \forall j \in J$$

$$X_i \in \{0,1\}; \forall i \in I$$

Second Model

$$\text{Minimize} \qquad \sum_{i \in I} \sum_{j \in J} dist_{i,j} d_j Y_{i,j}$$

Subject To:

$$\sum_{i \in I} \sum_{j \in J} (dist_{i,j} > HighServiceDist ? 0 : 1) d_j Y_{i,j} \geq HighServiceDemand$$

$$Y_{i,j} \leq (dist_{i,j} > MaximumDist ? 0 : 1); \forall i \in I, \forall j \in J$$

$$\sum_{i \in I} Y_{i,j} = 1; \forall j \in J$$

$$\sum_{i \in I} X_i = P$$

$$Y_{i,j} \leq X_i; \forall i \in I, \forall j \in J$$

$$Y_{i,j} \in \{0,1\}; \forall i \in I, \forall j \in J$$

$$X_i \in \{0,1\}; \forall i \in I$$

For both of these models, we have incorporated the *MaximumDist* family of constraints to ensure that no assignment exceeds a reasonable limit. The second model incorporates a new constraint using the *HighServiceDemand* constant. This value for this constraint can be determined from the result of the first model. For example, if the first model discovers a solution that meets 75 million units of demand with high service, then the second model can set *HighServiceDemand* to be 75 million.

Service-Level Constraints

Although the decision for Chen's Cosmetics was simply based on proximity to customer demand alone, many studies will want to consider service level as just a part of their entire network optimization. In this case, service level is no longer the overall solver objective but a constraint within the larger model.

As previously discussed, our first service-level objective asks the model to locate plants as close to as much demand as possible. A solution to this model can result in some customer locations being as close as 20km from the servicing plant while others are as far away as 4,000km. A modeler on Chen's team thinks it will be best for business if the solution ensures that *all* customers can be serviced within one-day transit or 400km. After quick analysis of their data, it is clear that this constraint will cause infeasibility in their model. In the table he created, as well as just a cursory glance at our map of the network including all potential locations shown in Figure 4.9, we can easily see that there are customer locations that are located more than 400km from any potential plant. Therefore, there is no possible way the solver can adhere to this new constraint.

Distributor Location	Distance to Closest Potential Plant
Kashi	1,076
Xining	694
Aksu	669
Kunming	632
Haibowan	536
Yinchuan	521
Lanzhou	514
Panzhihua	511
Nanning	505
Yining	499
Putian	476
Beihai	464
Xiamen	464
Haikou	456
Fuzhou	444
Zhangzhou	433
Liuzhou	427

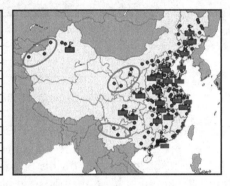

Figure 4.9   Customer Locations Farther Than 400km from Any Plant

Knowing this, the modeler now decides to alter the constraint. He realizes that the best that Chen's can ensure with their current set of potential plant locations is a maximum of three-day transit to all customers. In essence, he must now tell the solver that the plants selected *must* be in locations where *all* distributors are no more than 1,200km away. After this constraint is applied, however, the model quickly tells him his optimization run resulted in no feasible solution. He is slightly puzzled by this and must now begin to logically analyze the effect that adding this constraint has had on the model.

Infeasibility in modeling can be quite common. As we continue through this book, we will continue to introduce you to common methods of constraining model output in order to produce real and implementable solutions. Constraints within modeling are useful and necessary when applied well, but you will also find that the opportunities for creating unattainable and/or conflicting constraints can be limitless and must be considered with each additional constraint you apply. Chapter 12, "The Art of Modeling," expands on best practices in quickly finding and alleviating these infeasibilities, but service-level constraints are very commonly a main cause. Let's look further at how Chen's modeler determined the root of his infeasibility and the options for adjusting the model to adhere to the constraint within a feasible solution.

When infeasible solutions are determined, the most common way to pinpoint their cause is to test iterations of providing the model slightly more freedom in its decisions until a feasible solution is possible. In this case, Chen's modeler knows that he has applied only two major constraints in his model: (1) that only two additional plant locations may be selected for use in the solution and (2) that all distributors must be located within 1,200km from one of the three plant locations. He also knows that the model was feasible prior to adding the second constraint. Knowing this, he decides to start by running iterations of the model with increases in the maximum distance constraint. You can see the results he found in Figure 4.10.

Max Distance Constraint	Feasible?
1,200km - Original Constraint	No
2,000km	Yes
1,800km	Yes
1,600km	No

Figure 4.10   Max Distance Model Iteration Results

Based on the results of his test, he now knows that the infeasibility occurs when he asks the model to select plant locations that result in the ability to service all customers within a max of 1,600km or less. Based on his previous discovery that all customers are located within 1,076km of at least one potential plant, he can now also deduce that it's the combination of his constraints that causes the infeasibility. That is, even though all customers are located within 1,076km of a single potential plant, when we limit the model to select a total of only three plant locations, there is no feasible combination of three locations that fall within 1,200km of all distributors. He does one final test on the model and increases the max number of plants that may be selected to four and reruns. The final result is feasible, confirming that it is the combination of the number of plants and the distance restriction that is causing the problem.

There may still be a way for Chen's modeler to include his constraint within the model but avoid the infeasibility. Constraints are often altered to incorporate only a portion of the population of the model. In this instance, the model may be altered to constrain the resultant service level so that at least 75% of customers are within 1,200km. This way the modeler doesn't have to be absolutely sure that a constraint is feasible over an entire model. Looking deeper at model results of this type is also a quick way to pinpoint the cases where this constraint is not feasible (these instances will always fall into the 25%

of customers served from plants located farther away than 1,200km). Another way to adjust this constraint may be to create subgroupings of the customers based on their importance to Chen's Cosmetics. This might result in Tier 1, Tier 2, and so on, designations applied to each. Then the modeler may either apply different levels of constraints to each or apply constraints to only some of the subgroupings. In this instance, Chen's may only want to ensure that their Tier 1 customers are within a maximum of 1,200km but the model has no restrictions on servicing the other locations.

Although, in the end, Chen's team still decided to go with their original decision on plant locations, the team benefited greatly from the understandings that resulted from this analysis and also now understand a good deal more about applying service levels as constraints within a model. They also learned a major lesson on how to approach infeasible solutions and how debugging can lead to a much deeper insight into understanding what drives the model solutions as well as producing a feasible version of the model.

The Importance of Sensitivity Analysis on Any Solution

In this chapter we have not only learned a good deal about the inclusion of service levels in a model but also learned the value of running multiple scenarios to test and further understand the output and different potential "good" solutions. Network design models may make assumptions in many areas:

- Demand in the future
- Transportation costs
- Labor costs

Although intricate forecasting techniques and other detailed methods of data analysis will help modelers to create the best assumptions in any model, it is important that any and all solutions are tested for robustness. This practice is commonly known as sensitivity analysis. This practice arms the modeler with a better knowledge of what drives the model and answers. By testing changes in key variables such as demand, transportation costs, and labor costs, a modeler is ensuring that an assumption made in one model input is not dramatically altering the resultant savings and network recommendations within the solution.

Why Dual Variables (or Shadow Prices) May Not Be Valuable in Network Design

At this point, someone with a background in linear programming might be asking why we do not do the automatic sensitivity with linear programming. If you run a linear program, you get information on the sensitivity of the solution as part of the output. You may have heard this called "dual prices," "dual variables," or "shadow prices." That is, the linear program can tell you the value of each additional unit of capacity. Although this can be helpful in some cases, we do not think it is practical for the following reasons:

- It works only for linear programs and almost every network design model has integer variables. The integer variables can be as simple as forcing a customer to receive all its product from one warehouse.

- If you do have a linear program, this free sensitivity analysis is really valid only for small changes in the value. It does not give insight into large changes.

- With commercial network design software, you can get good reports on which constraints are tight, and this helps point you in the correct direction and gives very similar information to dual variables. That is, if a constraint is tight, you are likely to get a better answer by relaxing that constraint.

- We have found that with the easy user interfaces, powerful computers and solvers, and with the need to test solutions anyway, the best approach is to just run different what-if scenarios. This can give you insight into the value of changing constraints and parameters.

Let's look at an example of sensitivity analysis performed on Chen's Cosmetics new distribution structure in Europe. For demand numbers, Chen's modeling team uses the estimates from the marketing team that suggest they will be able to control 65% of the affordable-cosmetics market in Southwest Europe (Portugal, Spain, France, Italy, Croatia, and Switzerland). Although Chun Chen has a lot of trust in her marketing department, she, like all good managers, knows that a forecast is never accurate. And that forecasting becomes even more difficult in new markets. So Chun asks the modeling team to run some sensitivity analysis around the effect this forecast or assumed market share has on the optimal warehouse location solution. This way she can rest easy, knowing that the company's decision will remain optimal despite the possibility of their underperforming in comparison to these optimistic market share goals.

The team decides to start their analysis by running the original model scenario with 15% less demand than was originally assumed for their Southwest European region. The solution recommendation is the same as for that of the original scenario. Optimal warehouse locations remain in Leipzig, Marseille, and Odessa. This gives Chen's the assurance that their optimal solution is robust enough to withstand up to 15% less demand in the Eastern European market.

The team then continues to test the robustness of their solution by running scenarios with between 20% and 40% less demand at 5% intervals. You can quickly see in their solution comparison report, shown in Figure 4.11, that at 20% less demand, the solution selects three completely new warehouse locations in Bucharest, Milan, and Voronezh.

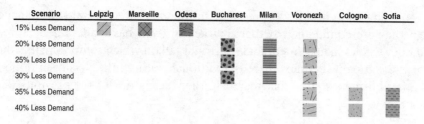

Figure 4.11 Demand Sensitivity Scenarios Warehouse Selection Comparison

By comparing the two solution maps in Figure 4.12, we see that the new location in the southwest, Milan, is located very close to the previously selected Marseille location; however, the territory it services has become much larger. And the two additional warehouse locations in this solution are in different regions farther east than Leipzig and Odessa. This solution makes complete sense in our "service based" optimization model. With less demand in the southwest, the model elects to move warehouses west to focus more on those steady and now relatively larger demand areas.

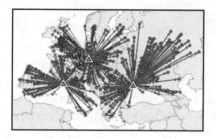
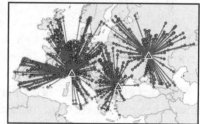

Figure 4.12 Original Solution Compared to 20%-Less-Demand Solution

With 35% less demand in the southwest, we see an even more dramatic shift. When the 20%-less-demand and 35%-less-demand solutions are compared, as shown in Figure 4.13, we see that the model still selects relatively the same locations to the east; however, it finds even less value with being located toward the southwest and elects to place the third warehouse location much farther north, in Cologne.

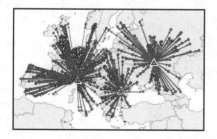
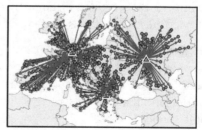

Figure 4.13 20%-Less-Demand Solution Compared to 35%-Less-Demand Solution

Chen's modeling team decides to present a summary of these results to management. Although they are confident that their marketing team has made a good forecast of demand, they will let management make the final call on which solution makes the most sense for the company. Perhaps there is additional research that has been done in this area or other business factors that may help them select one of the previous solutions over the other.

Whatever the final decision turns out to be, the modeling team has supplied exactly what network design studies are meant to do. These types of studies and the commercial applications designed to facilitate them are not meant to "make network design decisions" but to *support* a business's decision with as many scenarios and information on key drivers as is needed for them to feel confident they are making good strategic plans for the future.

Lessons Learned from Alternative Service Levels and Sensitivity Analysis Modeling

When we were modeling just the weighted-average distance in the previous chapters, we saw how the solution located points close to demand centers. When maximizing the demand within a certain radius, the solution didn't need to get as close to big demand centers. In fact, the solution pulled facilities to locations where they could get to more demand. That is, as long as the big demand points were within the radius, it did not matter how far away they were from the facility. So different measures of service can give you different answers.

In this chapter we learned that the term *service level* can mean many different things. For network design, we are concerned only with definitions that impact the location of our facilities. Even with this restriction, we learned about several definitions that apply to network design.

Modeling Service Levels as Model Objectives:

1. Minimize the average distance: Selects optimal network structure in order to minimize the weighted distance traveled or, simply stated, asks the model to service the highest levels of demand within the shortest distances.

2. Maximize the percentage of customer or demand within a certain distance: Maximizes customers or demand serviced within a maximum of x distance or, simply stated, asks the model to maximize the number of customers or percent of demand that can be serviced within a distance determined to be an acceptable service level.

Modeling Service Levels as Constraints:

3. Maximum Average Distance Constraints: Helps guide the model to return solutions with reasonable assignments for those customers that cannot be reached within a pre-defined service territory.

4. Maximum Distance Constraints: Forces the model to service all customers within a specified distance.

5. Maximum Distance Constraints per Customer Subsets: Tailors the previous Max Distance Constraint to a specific subset of customers within the model.

By adding constraints, we were also able to create situations in which the model was infeasible. That is, we gave the optimization model contradictory constraints. This helps us understand that we need to be careful with our model formulation so we return reasonable results.

Each of the three previous chapters has shown analysis by running several scenarios or versions of your models in order to understand and find our best solutions. Our discussions around service levels and the modeling we did for Chen's Cosmetics only further supported this practice. This led us to formally define this practice and its use in something called sensitivity analysis.

We now know that sensitivity analysis is the cornerstone for producing robust solutions that companies can depend on. This lesson becomes one of the most important lessons this book will cover. Even after finding the best solution, you, as a modeler, need to ensure the robustness of this solution. You should be confident in all your results and have a good idea of how all of your model variables are contributing. This is what moves you from someone who can build a model to an experienced supply chain network analyst!

End-of-Chapter Questions

1. A major consumer products company is making a major push to improve the number of orders filled on time in full. This measure is often abbreviated as OTIF. It means that when a customer places an order for several different products, you are able to ship all the products requested in the quantities requested. The company wants you to run a network design model with this as the sole objective. Why is a network design study not going to help this company?

2. A major chemical company operates a private fleet of tank trucks that deliver full loads of liquid chemicals from warehouses to all associated customers. The "customers" in this case are plants using these chemicals within production processes that run 24 hours a day. If these plants run out of any of the necessary raw chemicals, they must shut down the ongoing production, which wastes precious time and costs them a lot of money. For this chemical company to fulfill the needs of these

demanding customers, they need to offer no less than two-day service for all demand. The modeling team understands this policy but is puzzled about how they convert the service promise into a constraint within their network design model.

  a. Based on the following facts, show these modelers how to properly calculate the maximum distance (in miles) they should apply to constrain the outbound lanes in their model. (Your answer should be in the form of a formula including the solution.)

  Drivers can work for only 10 hours per day (this includes the prep time).

  Trucks travel at an average speed of 50 mph (miles per hour) when making customer deliveries.

  Tank trucks must be hooked up with hoses that load and unload the liquids. It takes approximately 1½ hours to fully load and unload one tank truck.

  b. What if the company decided to start using team drivers as opposed to the single drivers we assumed within our initial calculation? In this case, two drivers are assigned to each truck, allowing the team to work twice as long (20 hours as opposed to the original 10 hours) before having to stop for the day. Based on this change, what is the new mileage constraint that should be applied within their model?

3. If you are deciding where to position ambulances to provide emergency service to residents within a city, would you rather minimize the average time to each resident or maximize the percentage of residents within a certain time? Why?

4. If the mathematical formulation only had an objective that maximized the amount of demand within a certain limit (say, 800km) and no other objective or restriction on the distance, how would customers outside of the limit be assigned to warehouses?

5. "Service Level Sensitivity to Number of Plants Selected—Mini Case Study"

  You can access and download the model as well as detailed instructions for further analysis within the `Chen's End of Chapter Questions.zip` file on the book Web site. You will be asked to create and review Chen's Cosmetics' supply chain network in China, as well as prepare and run scenarios in order to determine the sensitivity of the overall service level that can be offered by allowing the model to select more plant locations.

  a. What locations are selected and what is the volume produced by each location if you allow the model to select a total of four plants? Five plants? Six plants?

  b. How much does service level (based on kilometers) improve with the selection of four plants? Five plants? Six plants? Give your answer in terms of kilometers and on a percentage basis.

5

ADDING CAPACITY TO THE MODEL

In the previous models, we did not consider capacity. Each facility could be as large as it needed to be to service the demand assigned to it. This may not always be a good assumption. In this chapter, we want to consider how capacity impacts our network design models.

Even though every facility in a supply chain has capacity limitations, you often do not need to include those limitations in the model. The main reason to omit capacity from your model is that it doesn't impact the decision you are trying to make. The following are reasons that capacity may not impact your decision.

First, when you ask the model to open multiple facilities, it will naturally tend to balance the demand across the facilities to minimize average distance or cost.

Second, especially for warehouses, increasing capacity can be relatively inexpensive. With warehouses, you may be using a third-party warehouse and extra space simply means asking for it. Even when you are not using a third party, many warehouses can easily expand in space. For manufacturing sites, extra capacity may simply mean hiring more workers (which you would have to do anywhere in the supply chain). Of course, the exception to both of these cases is when you have highly automated facilities with expensive equipment.

Third, often the decisions you are making concern the long term. And, in that case, today's capacity is not fixed for the future. You want the model to return the size of the facility needed. You do not want to fix this beforehand.

However, there are instances when capacity will be critical to your network design study. For example, if you are making decisions about investing in expensive, highly automated plants or warehouses, or are deciding where to place a specialized production line, or trying to determine how many shifts you want a plant to operate.

Following are some short cases that help highlight some issues you may find when modeling capacity.

Case Study: Swimming Pool Chemicals

We worked with a company that made and sold chemicals for swimming pools. Because they could never exactly predict where and when the first warm weekend would hit and everyone would want to buy supplies for their swimming pools, they wanted to position their product in warehouses that were as close to the customers as possible. However, they also wanted to minimize the number of warehouses so that they had some flexibility. For example, if they ship product to a region and that region does not get warm weather, they may have to ship this product out to another warehouse. They did not have the luxury of being able to afford these extra shipments. So they wanted to avoid it, if possible.

However, they had one other thing to consider: Their products were labeled as hazardous, and if the warehouse where they were being stored caught fire, the fire would be difficult to contain and would strain the local firefighting squads. Therefore, the local fire departments would not allow them to have more than 100,000 units of product in the warehouse.

So when the company designed their supply chain, they had to consider this capacity constraint. This could change their answer. For example, without capacity constraints it might be best to locate very large facilities near densely populated areas and smaller facilities in other locations. But with this constraint there will tend to be many warehouses at or near the limit of 100,000 units. And, some densely populated markets may have multiple warehouses whereas without the constraint one large warehouse would have been better.

Case Study: Warehouse Capacity Utilization

Sometimes, the use of math in business is not as precise as the use of math in engineering.

When an engineer expresses the capacity of a warehouse, it is the maximum amount of product that can be stored or processed. Here we are talking about the true maximum amount of capacity. Beyond this point, the firm has to rent extra space or take some other measure. At most, a warehouse can be at 100% of capacity. This percentage is the utilization of the warehouse.

We have been involved in many discussions with managers of warehousing or directors of the supply chain where we have heard a quote like this: "Our current utilization of warehouse A is 125% and warehouse B is 133%." And the manager will stick to these numbers even after you qualify that he is not measuring against some theoretical capacity or a desired level of capacity.

If you are thinking like an engineer, a statement like this can hurt your brain. How can they be running the warehouse at more than capacity? It turns out that warehouse capacity is very hard to measure, and it may not even be possible to come up with a single number that captures it.

Warehouses have three primary functions. They receive product, they ship product, and they store product.

The capacity for shipping and receiving depends on how many shifts you run, how many people or resources (like automated pickers) you have, what kind of equipment you have, and how much space you have dedicated to these functions. In highly automated systems, it may be easier to measure this capacity. In manual systems, you can easily add more people. But as you add people, you increase the congestion in the warehouse and that may lead to less productivity.

The storage capacity is measured by how much physical space you have and how much is in the warehouse at any given time. However, this is not easy to measure. For example, a warehouse may have storage areas where items (mostly in pallets) sit on the floor and areas where items are stored on racks. These warehouses need to leave room for aisles so that you can get to the product. Also, the managers of these warehouses might like to put the same products in the same location so that they can find them. You can understand the difficulty in measuring capacity when the warehouse starts to get full. Most warehouse managers can get creative when they start to run out of space. They can stack items (by stacking pallets or adding racks to the floor storage areas), place the same product in different locations, use the aisles, and not unload incoming trailers (a common trick).

So a warehouse capacity utilization of over 100% is not violating some rule of mathematics. Instead, from a business point of view, it means that the warehouse is very busy. We don't know whether it is so busy that it is inefficient or whether something should be done about it. That is left for further analysis.

These cases do, however, cause you to pause when modeling warehouse capacity. If a warehouse can handle 120% of capacity, you might not want the model to make different decisions when it hits 100%. Maybe you should include some slack in your measure of flexibility. We'll have more on this topic later.

Case Study: Paint Company and Capacity

Manufacturing capacity is typically more straightforward to measure than warehousing capacity. And it is usually more expensive to add. Of course, there are some difficulties because the product mix may not be known in advance, the number of setups can impact capacity, and there are always unanticipated problems on the shop floor.

However, these problems can usually be accounted for in network design without too much effort. For our purposes in locating facilities, we can determine how much product can be produced over the year for a given number of shifts. This measure takes into account the product mix (by directly accounting for capacity needed per product), the average amount of time lost to setups, and the average uptime on the lines.

In some cases, hitting a capacity constraint can have serious implications for the business. We worked with a regional U.S. paint company that had two plants—one on the East Coast south of New York City and another in the Southeast near Charlotte, North Carolina. They sold through their own retail stores in about 30 major metro areas east of the Mississippi.

When we first started working with the company, they distributed their product directly from warehouses that were attached to their plants. Because each plant did not make the full assortment of products, they had to ship a lot of product between the two plants so that they could ship a full assortment to their stores.

Paint is very expensive to ship, and they were always looking for ways to reduce transportation costs. So, for example, they considered the following questions:

- Which warehouse should serve which metro area to reduce transfer of paint between the two plant warehouses?

- Should they add a warehouse or two to get closer to their stores and minimize the shipments between the two plant warehouses?

It turned out that a new warehouse to serve the metro areas around Chicago and Minneapolis could reduce transportation costs. However, the company recognized that they also had to design their supply chain for their future growth. Both of their plants were running close to capacity. If they continued to grow at the current rate, they would have to expand one or both of their plants.

By adding future demand to the model and not allowing any expansion, they let the model make a key decision: which markets they should exit. As a result of the analysis, they sold their stores in the Chicago and Minneapolis areas to another paint company, delayed a significant investment in their plant capacity, and reduced their per-unit transportation costs because they were much closer to their demand.

This case clearly highlights the trade-offs when you bump up against true capacity constraints—either you have to add capacity or you have to not meet all your demand. Network modeling can help you make decisions like this.

Adding Capacity to the Model

The previous cases highlight how capacity can play a role in your model and the care you should exercise when adding capacity to your models.

For example, when modeling warehouse capacity, you have to understand whether the capacity measure is a hard constraint or whether there is some slack in the calculation.

If there is some slack, adding capacity to the model may mean simply increasing the number you were given to better reflect reality. In other cases, managers will recognize that they cannot really run a facility at 100% capacity. Instead, they want to say that the limit is 80%. You should note that this does not solve our problem of whether this is a hard constraint. Is the 80% the hard constraint or is the 100%?

If the warehouse constraint is real, you need to consider the cost of new warehouse space. If you are using third-party space, adding capacity may just incur extra variable costs. Or if your warehouse systems are relatively simple and manual, adding space may be no problem at all. For example, if you are currently storing product directly on the floor and your ceilings are high enough, you may simply need to add racks to increase the storage capacity. If you need to have a high-speed, highly automated warehouse, you will need to think about a more significant capital investment.

If the warehouse constraint is not in space, but in terms of its ability to process inbound and outbound shipments, increasing capacity may require additional shifts, additional dock doors, or more automation.

With plants, capacity expansion typically comes in three forms:

1. **Adding Labor**—Extra labor can provide the ability to produce more. Sometimes extra labor can be added in small increments (1 person at a time, for example) or in larger batches (you need to bring in another full crew).

2. **Adding Shifts**—When you add shifts, you may incur additional fixed costs (to staff up the line, to manage it, and it may be difficult to remove later), as well as additional variable costs.

3. **Adding Equipment**—This can include anything from investing in the existing equipment to make it faster, to adding production lines, to building a completely new plant.

With both plant and warehouse capacity, you may also need to factor in the time period for capacity. For example, in annual or long-term models, you will model the effective capacity over the year. If you have a very seasonal business, you can adjust the capacity to reflect that fact. For example, you can set your warehouses to average capacity or peak capacity, and you can do the same for plants.

Mathematical Formulation

Initially, we will add only constraints to warehouse capacity. We will assume unbounded production capacity in the model we formulate next.

$$\text{Minimize} \quad \sum_{i \in I} \sum_{j \in J} dist_{i,j} d_j Y_{i,j}$$

Subject to:

$$\sum_{i \in I} Y_{i,j} = 1; \forall j \in J$$

$$\sum_{i \in I} X_i = P$$

$$\sum_{j \in J} vol_{i,j} Y_{i,j} \leq cap_i X_i; \forall i \in I$$

$$Y_{i,j} \leq X_i; \forall i \in I, \forall j \in J$$

$$Y_{i,j} \in \{0,1\}; \forall i \in I, \forall j \in J$$

$$X_i \in \{0,1\}; \forall i \in I$$

This is identical to the "distance only" problem we formulated in Chapter 3, "Locating Facilities Using a Distance-Based Approach," with the following changes:

- We create the term cap_i to measure the capacity of warehouse i. For our purposes here, we have a single measure of capacity and just on the warehouses. This will serve our purposes here of giving you the intuition needed for thinking about capacity. In practice, we model capacity in many ways. At a warehouse we measure it as a storage capacity, a total throughput capacity, or an ability to process inbound or outbound shipments. At a plant, likewise, we can measure capacity by the overall time available in regular time and overtime, the overall tons or units that can be produced, or the maximum amount of a product you can produce. There are many ways to capture capacity at a plant, but because this is effectively the same as for a warehouse, we won't cover it here.

- We create the term $vol_{i,j}$ to measure the effective volume of demand for customer j being assigned to warehouse i. Note that we index this term by both the customer and the warehouse. This allows us to model the possibility that different warehouses might use different units of measurement to determine their capacity, or that different products might be cross-docked with higher levels of efficiency at some facilities. When this term applies to a plant, it has the same definition: It is the amount of a plant's capacity that is consumed by that customer's demand. Again, a more detailed version can allow for the capacity consumed to vary with the product as well.

- We add the constraint $\sum_{j \in J} vol_{i,j} Y_{i,j} \leq cap_i X_i$ which we apply to all warehouses. This constraint ensures that a warehouse is never assigned more demand than it can handle. Note that if the warehouse capacity is not infinite, this constraint also will ensure that if a customer is assigned to a warehouse, that warehouse must be opened. However, we will not discard the family of $Y_{i,j} \leq X_i$

constraints, as experience has shown that their inclusion can significantly improve the runtime of our models. (The engineering of optimization problems often involves applying potentially redundant constraints in order to generate efficient runtimes. This phenomenon is but one such argument in favor of purchasing off-the-shelf optimization applications for specific problems, as opposed to using generic tools to create optimization problems by hand.)

Possible Difficulty with Models That Have Capacity Constraints

Like our service-level constraints in the previous chapter, we need now to be careful that the problem is still feasible. That is, it is again possible to create a model such that it is impossible to solve. For example, we can specify 100 units of demand and only 80 units of capacity. In this case, the constraint that we must meet all demand cannot be met. We need to make sure that our constraint on the number of sites does not conflict with the capacity constraint. There are also ways to set up the model so that you do not have to meet all demand, but just the most profitable.

Depending on how you set up your model, this capacity constraint may be just another simple constraint that pushes your model in a new direction with no impact on runtime. However, it is also easy to set up a model using capacity constraints that becomes nearly intractable. When a model with capacity becomes intractable, it is usually because the capacity required of individual customers is large relative to the overall capacity, and you are forcing all demand for each customer to be serviced from only one location (often termed single sourcing).

For example, say that you are locating two warehouses, each with 100 units of capacity, and you have three customers, each with a capacity requirement of 60 units. System-wide there is enough capacity (200 units); however, no two customers will fit into one warehouse (two customers need 120 units but a warehouse has room for just 100).

Why can this happen?

When you are locating warehouses, you may want to specify that certain customer groups are assigned to a single warehouse. That is, all your customers in Los Angeles should be assigned to a single warehouse. As orders come in from Los Angeles customers, you don't have the capability to try to figure out how to allocate these orders to more than one warehouse. As another example, a retailer may want to group their stores based on the Sunday-paper coverage. That is, promotions may run in the Sunday papers, and you want all the stores in one region to be served by a single warehouse to better manage the promotions.

In the home-building industry for products such as roof shingles, wood boards, or siding, the exact color of the product may vary ever so slightly depending on where the product is made (depending on the trees grown in that area or the type of soil in that area). This becomes a problem (and quickly becomes a major problem) when a customer uses your product from two plants on his house and the color difference is quite noticeable. Customers do not like this. Contractors do not want to get into this problem. So, to avoid this problem, these plants will service large areas where contractors are unlikely to get product from multiple locations. In this case, you may want to assign all the customers in California to a single plant.

When modeling manufacturing capacity (and especially when doing it in weekly or monthly buckets), you may want to specify minimum quantities to run (often called "lot sizes"). So when you run a product, you want to make a lot of it.

In all three cases, we are assigning capacity in large chunks. In the first case, we cannot split the demand of Los Angeles—it is either assigned to a warehouse or not.

If, in any of these cases, the capacity used by these chunks is large relative to the total capacity, then what we have done is incorporated some elements of one of the classic problems in computer science—the "knapsack problem." This problem tries to allocate discrete resources in a way that maximizes total value, while staying below a total limit on capacity. For example, if you are packing for a long trip, you would want to bring the most important collection of items, without exceeding the capacity of your suitcase.

Unfortunately, knapsack problems can be hard to solve. And knapsack problems embedded into our already very large network design model (think back to the number of combinations in earlier chapters) can cause serious problems with the runtime of the models.

If you imagine an extreme version of our model with just two warehouses, one cheap and nearby, and the other expensive and quite distant, with a small set of customers that each consume a relatively large percentage of that available capacity, then our network design problem would be equivalent to a knapsack problem. We would be trying to assign as many customers as possible to the close warehouse, while using its limited capacity as efficiently as possible.

Intuitively, it is easy to see why such a knapsack problem can be very hard. Although the temptation might be to greedily assign the biggest, most important customers to the nearby warehouse, this might result in a configuration in which this warehouse isn't used to its full capacity. The optimal set of assignments might be ones that don't make much sense at first glance, as a far-sighted solver might choose to localize oddly sized customers with a relatively small incentive to use the nearby warehouse in order to create a "just right" fitting of the limited space. (See sidebar for more information on the difficulty of these problems.)

When you set up capacity constraints, you should watch out for inadvertently embedding a hard knapsack problem into your model. If your capacity is measured in large chunks, you are embedding a knapsack problem into an already difficult problem. This causes three problems that may not be intuitive:

1. The problem is infeasible and in a way that is not clear. You can easily have a case in which the total capacity is more than enough for the demand, but because of the way demand is grouped, it does not all fit into the capacity.

2. The solution looks bad. The optimization returns a solution that does not look intuitive. Customers are not assigned to the closest warehouse and may be assigned to warehouses quite far away. The model has found a way to meet all the demand, but it had to do it through a clever combination of assignments— as if it were solving a big puzzle.

3. The runtimes are very long. We already established the difficulty of a knapsack problem on its own. Now, we are also asking the optimization to pick the best number of warehouses out of a given set. Combining these two hard problems is worse than just the sum in terms of complexity and therefore takes much longer to solve.

A Short Note on Computational Complexity

It is somewhat challenging for a book of this sort to present an appropriate amount of theoretical background. On the one hand, the target audience for this book does not have a pressing need to understand the theories of computational run times that have been generated by mathematically inclined computer scientists over the past 40 years. On the other hand, real-life users of strategic network design solutions are often surprised by the practical implications of such abstract underpinnings. Slight alterations to the input data of such models can result in significant increases in the time required to generate a high-quality solution. Even worse, naive users can sometimes fall prey to the idea that there are foolproof fixes to such difficulties. In fact, strategic network design problems go to the heart of the most challenging problems in industrial computing, and it is worthwhile to build some intuition as to why their solvers cannot be guaranteed to run quickly on all problem instances.

The key concept to algorithmic classification of problems is that of reduction. One problem can be reduced to another if the results of the second problem can be used to easily solve the first problem. In cooking, the problem of making spaghetti can be reduced to the problem of boiling water. After you have boiling water on hand, preparing spaghetti is fairly easy.

With computational reduction, the second problem is often thought of as being a useful subroutine for a programmer trying to solve the first problem. Theoretical computer scientists will demonstrate that A reduces to B by writing the outline of a computer program that solves A by exploiting a black-box subroutine that can efficiently solve B. Of course, such exercises are academic, as there need not be any solver engine that effectively addresses B. These programs are thus rarely coded, and their accuracy is verified by the examination of other computer scientists.

If one is already confident that problem A is very, very hard, then a demonstration that A reduces to B is a de facto recognition that B is also very, very hard. If this sounds odd to you, consider the converse. Suppose B wasn't hard at all. That is, suppose there exists an engine that can solve problems of type B very quickly. This means that A must not be very, very hard, because we can solve A simply by executing a computer program that uses B as a subroutine in order to address A.

Although this sort of reasoning does have a sort of whiff of fatalism to it ("Huzzah, A reduces to B, so we can give up on trying to quickly solve B"), it can also be used in some very practical applications. In general, the easiest way to determine that a problem is in fact hard is for this problem to play the "B" role in an "A to B reduction," where A is already known to be hard. This strategy can then be repeated to corral a stable of effectively unsolvable problems. We can then recruit the most convenient character from this rogue's gallery to construct a "hacker-resistant" barrier for information that needs to be kept secure.

For example, the RSA cryptography algorithm is based around the idea that integer factorization is very, very hard. That is to say, given a very large integer, there is no easy way to determine the prime factors that would be multiplied together to generate this integer. A hacker, when attempting to decrypt a message encoded with RSA encryption, would be implicitly confronted with a very large integer factorization problem. Thus, the known difficulty of integer factorization can be used to allow for the creation of a trillion-dollar industry (e-commerce) by allowing people to trust that their credit card numbers cannot be easily read by unseen Internet hackers.

However, integer factorization isn't the only problem that is known to be very, very hard. There is a mathematical problem called Fixed Charge Network Flow. FCNF problems look very much like strategic network design problems. They involve sites (usually called nodes) and transit lanes (usually called arcs). Some nodes are capable of producing goods, and others are obligated to consume goods. A given arc will charge a one-time cost penalty to be "opened for business," and another cost penalty based on the quantity of goods it carries. An FCNF engine (reliably efficient or not) will determine the least expensive way to transfer the correct quantity of goods from the supply nodes to the consumer nodes.

As you might expect, one could solve a wide variety of strategic network flow problems with an FCNF engine. What is less obvious, but no less real, is that you could also solve integer factorization problems with such an engine. Were a brilliant computer scientist to someday deduce an algorithm that could be used to quickly solve all manner of FCNF problems, the world of e-commerce (and indeed, of computer security in general) would be brought to its knees. (The 1992 movie *Sneakers* explores this idea, albeit as part of a somewhat far-fetched conspiracy-theory tale.) This is not to say that computer scientists are not trying to create efficient FCNF engines. However, the known difficulty of this type of problem gives little hope that they will ever create a truly efficient strategy for solving all FCNF problems.

This broader lesson is true for strategic network design problems as well. If there were some grand breakthrough in addressing such problems, its repercussions would be felt far outside the realm of supply chain planning. You would be informed of such a discovery on the front page of the newspaper, not within the niche area of network design. The computational models we are using have a utility far beyond our scope, and significant advances in their tractability would quickly reverberate throughout modern society.

Although supply chain planners are fortunate in that they often formulate problems that can be solved quickly, they should be aware that there is no guarantee that all of their models can be addressed in an efficient manner. For example, if a hard knapsack-like problem gets embedded in your model, you may not be able to generate an answer. However, the planner may perhaps take some comfort in the knowledge that their RSA-encrypted credit card number is safe and sound.

How Capacity Constraints Can Change a Model

It is important to understand how capacity constraints change the solution of a model.

Let's start with an example of a distributor in Brazil that services 25 customer regions. Each region has a different demand. The total demand is approximately 100 million units. We would like to locate five warehouses to service this demand. If we run the model without capacity constraints and ask for the best five warehouses, we are provided with the solution shown in Figure 5.1. We are specifically interested in the location and throughput of each warehouse selected.

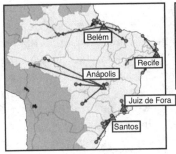

Warehouse	Throughput
Santos	41,545,912
Juiz de Fora	23,930,340
Recife	17,936,334
Anápolis	7,986,478
Belém	7,267,529

Figure 5.1   No Capacity Model Output

The throughput per warehouse ranges from only 7.3 million units all the way up to 41.5 million units. This clearly shows us that the customers along the southeastern coast have the majority of the demand and that placing warehouses close to them is a major driver for minimizing weighted distance traveled for all deliveries.

Now, let's apply a capacity constraint in an attempt to more evenly balance the throughput of each facility. To do this, we can simply allot a capacity of one-fifth of the total model throughput to each warehouse. In this case that means that every warehouse in the model has a maximum capacity of 20 million units (approximately 100 million units/5 warehouses we are allowing the model to select). We will review two scenarios to understand the impact this will have. In the first case, we will run the model forcing the use of the same five warehouses selected previously and show the effect this has on our solution.

After attempting to run this model setup, however, we find that no feasible solution exists. There are two ways this model may be infeasible. The most obvious is if we set the total system-wide capacity to a number less than the capacity required. However, we know that this isn't the case based on our previous calculations for overall throughput and per-warehouse capacity reviewed earlier. This leads us to our second case.

We quickly realize that adding this capacity has now caused contradicting constraints somewhere in our model. The 20-million-unit capacity constraint we added at each warehouse attempts to force each warehouse to hold almost exactly that amount with no extra room to use elsewhere as our total flow of goods is approximately 99 million units. Therefore, no warehouse will have space for more than 1 million additional units within the final solution. In addition, our model is set to "single source" all customers. This means that a customer may receive product from only one warehouse location. Considering these two constraints together, if customer demand cannot be equally split up in 20 million unit buckets (without splitting demand for any of the customers between groups), then this problem becomes infeasible. Taking a closer look at our model in Brazil, we find that the Sao Paulo Region customer actually has a total demand of approximately 29 million units. This discovery gives us a clear insight into our infeasibility. Not only do we have to ensure that we have enough capacity for the overall demand, but we also must ensure that we have enough capacity at single instances of the warehouses for specific sourcing rules (like single sourcing) that we may have applied.

If we increase our capacity constraint to 30 million units per warehouse, however, we can now compare the original no-capacity solution on the left to one with capacity on the right, as shown in Figure 5.2.

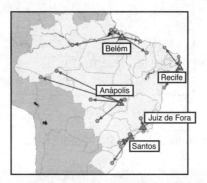

Avg Distance to Customer: 326kms   Avg Distance to Customer: 412kms

Figure 5.2   Capacity Scenarios Solution Comparison

You can quickly see the adjustments the model had to make in regard to which customers the Santos warehouse can still service within this capacity constraint and which must now be serviced by Anápolis or Juiz de Fora. As a result of this, the average distance that must be traveled from warehouse to customer increases from 326km in our original solution to 412km in this scenario as well.

If we review the output of throughput between both scenarios as shown in Figure 5.3, we see a slightly more balanced number of units serviced by warehouses in the southeast while those in the north remain unchanged. Their ability to service their originally assigned customer base was well under 30 million units and therefore no change is required.

Warehouse	Original Throughput	30 Million Unit Capacity Throughput
Santos	41,545,912	29,029,226
Juiz de Fora	23,930,340	28,842,504
Recife	17,936,334	17,936,334
Anápolis	7,986,478	15,591,000
Belém	7,267,529	7,267,529

Figure 5.3   Capacity Scenarios Solution Throughput Comparison

Let's now let the model select any five warehouses with this capacity constraint. We can find the resultant solution compared with our original solution in Figures 5.4 and 5.5.

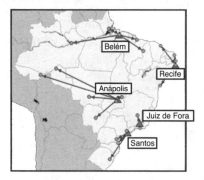
Avg Distance to Customer: 326kms

Avg Distance to Customer: 341kms

Figure 5.4   Original Versus Five-Warehouse Solution Map Comparison

Warehouse	Original Throughput	Best 5 With Capacity Throughput
Santos	41,545,912	29,029,226
Juiz de Fora	23,930,340	29,042,104
Recife	17,936,334	17,936,334
Anápolis	7,986,478	-
Ponta Grossa	-	14,712,066
Belém	7,267,529	7,946,863

Figure 5.5   Original Versus Five-Warehouse Throughput Comparison

What has changed in the solution? The number of facilities has remained the same, but the model has to place another warehouse closer to the coast to better fulfill the higher-demand areas there, which forces the removal of Anápolis, requiring much farther distances to travel to service both the central and the northwestern geographies. All the facilities are now relatively closer to the same size. But the average distance to customers has gone up. When adding constraints to any model, we should remember that the solution will never improve. And if the constraint is meaningful enough, the results of our objective may get much worse. In this case, the result proves to be approximately 5% worse.

Although constraints do ensure worse results, we must remember that they are sometimes a necessity in order to produce implementable results. The key to their use in modeling is to ensure that we are applying only meaningful constraints that aren't causing unneeded pressure on the model or, even worse, leading us to a problem with an infeasible solution.

Lessons Learned for Adding Capacity to Our Models

Capacity constraints don't necessarily change the locations of facilities, but they do have the impact of changing the warehouse to customer assignments. With capacity constraints, the assignments may look completely strange and seem to contradict the objective. In tightly constrained models, the optimization has to do everything possible just to find room in a facility, and only then can it worry about trying to minimize the weighted-average distance (or other objective).

One lesson about capacity is that even though every supply chain has capacity constraints, you do not always want to model them.

Capacity can be difficult to measure. When you include a capacity constraint, make sure that you model it carefully and that it is going to do what you want it to do.

If you add a capacity constraint and entities in the model consume capacity in large chunks, you may be creating a problem that is very hard to solve. In addition, capacity constraints can cause infeasible models. Remember that we also saw infeasible models in the preceding chapter on service levels. It is worth noting that when you combine both of these constraints in a model, you may be creating more possibilities for infeasibility as these constraints interact.

End-of-Chapter Questions

1. For this problem, let's consider a simplified version of the problem similar to the distributor in Brazil. The firm we are considering has three facilities, each with the capacity to serve 20 million units of demand. Assume that there are nine demand regions and that a demand region must be served by only one facility. If the demand (in millions) for the nine regions is 9, 7, 6, 6, 7, 6, 6, 7, and 6, explain why this problem has no feasible solution.

2. For the same problem as the preceding one, assume that the demand (in millions) for the nine regions is 9, 7, 6, 6, 7, 3, 10, 7, 5. What is a solution to this problem? When you are assigning regions to facilities, how much flexibility do you have to assign every region to the closest facility?

3. Assume that you have two manufacturing plants that need to make nine different products. Assume that annual demand for each product is 1 million units and that to achieve economies of scale you need to make at least 600,000 units at a single plant. You would also like to have each plant make the same number of units. What will happen to your model if you put in a constraint saying that plant capacity is 4.5 million units and at least 600,000 units of a product must be made at a plant?

4. Classic Linear Programming Transportation Problem I. The classic transportation problem that you can find in most linear programming books (or by a quick search of the Internet) is set up something like this:

   You have a set of source points with a given amount of supply available and a set of demand points with a required amount of demand needed. The total supply equals the total demand. However, the individual supply and demand points do not need to match up. For example, let's assume that Supply Point #1 has 100 units available, Supply Point #2 has 120 units available, and the three Demand Points require 75, 90, and 55 units. Then, there is a cost to assign each demand point to each supply point.

   a. Explain how this model is equivalent to the model in this chapter.

   b. What assumptions does the Transportation Problem make about assigning demand to source points? How is this different from some formulations we have made in this chapter? How would this impact the Classic Transportation Problem?

5. Classic Linear Programming Transportation Problem II—Mini Case Study. Open the file `LP Transportation Problem.zip`. This file contains more directions and the model. In this case, you have four coal mines and 15 power stations you are servicing. In this case, we are interested in minimizing the average distance traveled.

   a. Run the model and show which coal mine should serve which power station. What is the average distance of this solution?

   b. Now, increase each plant's capacity by 10% and rerun. How did the solution change? Did the average distance decrease? Why?

   c. Now, go back to the original model and force each power station to receive product from just one coal mine. Why is the model infeasible?

   d. Now, remove the capacity constraint on the coal mines and rerun. What is the new solution? What happened to the average distance? How much is being supplied by each coal mine? How is this different from the original supply constraints?

6. Brazil Model: Open the file `Brazil Capacity.zip`. This is the same model as highlighted in the case. Additional information and directions are in this file.

   For this assignment, go to the scenario with a capacity of 20 million per warehouse. Now, relax the constraint that forces every customer to be served by one warehouse. What are the best five warehouses? How does the average distance compare to the unconstrained and 30 million model? Why do you think the average distance changed? How much product does each demand region receive from each warehouse?

6

ADDING OUTBOUND TRANSPORTATION TO THE MODEL

Up to this point, we've looked at models that have focused mostly on service levels and locating facilities as close to demand as possible. However, running a supply chain costs money. And, sooner or later, someone is going to ask you to financially justify the new supply chain design. That is, you are going to have to determine how much money you will save with the changes you are suggesting.

Transportation is often the most important cost in a network design study. A retail or manufacturing company may spend the equivalent of 2% to 3% of their total revenues on transportation alone. For a company with $5 billion in revenues, this could equate to $150 million in spend. You can imagine that reducing these costs can make a significant difference to the firm's performance.

It is by design that we have selected transportation costs as the first cost category we discuss in this book. A change in the number and location of facilities often impacts transportation costs more than any other cost in the supply chain. Conversely, manufacturing, purchasing, or product handling costs may be fairly constant no matter where your facilities are. Therefore geographical changes to a network may have less impact on these costs.

In practice, as with service-level models, many network design studies consider only transportation costs. And these studies are perfectly valid. It is much more common to include just transportation costs when you are considering warehouse locations only. But it can happen with studies determining manufacturing locations as well.

To keep with our theme of creating building blocks for more complex models, we will analyze a simple model with a set of facilities that ship directly to customers. Often, this is called outbound transportation. That is, we are shipping "outbound" from our facilities. In later chapters, we will expand our analysis to also include the transportation costs to get product into the facility, from a plant to a warehouse, before it moves on to a customer. This leg of transportation is called inbound transportation. These terms can get confusing when you have multiple levels in the supply chain however. For example, you can have inbound shipping from the raw materials plant to the manufacturing plant before the manufacturing plant then ships product out to the warehouse. You may also have cases in which a product ships through several warehouses (central and regional) on its way to a customer. Despite this, we will stick with the industry-accepted terms throughout the book.

With this introduction of transportation costs into our modeling, we will still be answering the question about where to locate a given number of facilities, but our objective will be to minimize cost, not the average distance.

Sometimes the inclusion of transportation costs to a practical center of gravity model will not change the solution at all, which can be attributed to many factors including the high correlation between distance and transportation costs. It is a good modeling practice to be aware of variables that do not affect the solution. We will cover this further in Chapter 12, "The Art of Modeling." In many cases, the solution will change as you add transportation costs to the model. There are many different reasons for this:

- **Minimum costs in transportation**—When you ship a full truckload, there is often a minimum charge assessed independent of how far you go. That is, a carrier may charge $1.50 per mile per truckload, with a minimum of $300. Therefore any shipments within 200 miles will cost the same. This cost structure will cause the optimization to be indifferent when assigning customers to warehouses within 200 miles as the cost will be $300 no matter what.

- **Regional differences in transportation costs**—Transportation rates are not symmetrical due to the imbalance of supply and demand. That is, a carrier may charge $1.25 per mile to move a full truckload of product from Kentucky to Florida but may only charge $1.10 per mile to move a full truckload of product from Florida to Kentucky. These rate differences are mainly attributed to whether the carrier has demand for its services in the region the truckload is being delivered to. If the carrier can easily find other cargo in that region, the cost to ship into there will likely be cheaper than deliveries to regions where cargo is more sparse and unpredictable. By adding these types of transportation costs to the analysis, the optimization may be able to locate facilities in areas with more favorable rates.

- **Different customer shipment profiles**—In a single network some customers may always order a full truck's worth of products at a time, while others may only order a half a truck. When calculated on a per-unit basis, the full-truck customers' transport costs are much cheaper than the transport costs of those who order in half truck quantities. In these cases, the optimization has an incentive to select facility locations closer to the more expensive half-truck quantity customers.

Formulating and Solving the Problem

In the model formulation, we are now minimizing the total transportation spend by picking the best number of facilities:

$$\text{Minimize} \quad \sum_{i\in I}\sum_{j\in J} trans_{i,j}\, d_j Y_{i,j}$$

Subject To:

$$\sum_{i\in I} Y_{i,j} = 1; \forall j \in J$$

$$\sum_{i\in I} X_i = P$$

$$Y_{i,j} \leq X_i; \forall i \in I, \forall j \in J$$

$$Y_{i,j} \in \{0,1\}; \forall i \in I, \forall j \in J$$

$$X_i \in \{0,1\}; \forall i \in I$$

The only significant difference between this model formulation and the previous practical center of gravity formulation is that instead of a distance matrix, we have a cost matrix. The matrix $trans_{i,j}$ represents the cost to send one unit of demand from facility i to customer j. For a reminder of how to read this model, see Chapter 3, "Locating Facilities Using a Distance-Based Approach." As a summary, here is what this model is doing.

The *objective* of this problem is as follows:

- Minimize the total transportation costs from the facilities (plants or warehouses) to the customers.

We also have the same *constraints* or rules to keep in mind that we previously defined and discussed in terms of the practical center of gravity formulation:

- We have to meet all the demand. This one seems obvious, but we do not want to minimize the costs by not serving the customer.

- We are limiting the number of facilities to P. If we did not limit the number of facilities and transportation costs were proportional to distance, the optimization engine would likely elect to open and utilize every potential warehouse within the model.

We are making the following *decisions:*

- Do we use the warehouse at location *i*? We will denote this decision as X_i. If $X_i = 1$, then we use the facility at location *i*; if $X_i = 0$, then we do not use that facility. This is what is called a binary variable. It can take on a value of only 0 or 1.

- Does facility *i* service customer *j*? We will denote this decision as $Y_{i,j}$. If $Y_{i,j} = 1$, then facility *i* services customer *j*. If $Y_{i,j} = 0$, then facility *i* does not service customer *j*.

There are three things you might not have noticed about the transportation costs when you read the description of the objective function. First, the demand is expressed in total (usually annual) and not shipment by shipment. Second, the transportation costs are expressed in terms of a cost per unit. That is, we are multiplying transportation costs by units of demand, not by shipments. And, third, you need to fill a transportation cost for every possible source and destination, not just the source and destination shipments you used in the past.

We will explore each of these, in turn, in the next several sections.

Demand is Expressed in Total, Not Shipment by Shipment

In the underlying mathematical formulation, we are *not* modeling every single shipment. Our demand is expressed as a single value (most likely annual), not a series of shipments.

It may seem more natural to simply model every shipment. Although, in theory, we could try to set up a model with every shipment. In practice, this model would become unwieldy and introduces a false sense of accuracy.

It becomes unwieldy, because if we model every shipment, we have to create a model with potentially hundreds of thousands of shipment points instead of several hundred customer points. We may now have 100 or 1000 times the data to analyze and the size of the model will likely bog down our machine and render the model unsolvable.

But, more important, we have created a false sense of accuracy. As in our discussion in Chapter 1, "The Value of Supply Chain Network Design," adding every shipment certainly feels more precise. However, it is not more accurate. Remember, we are running models for the future—the next year, the next five years, or maybe even the next month. We are fooling ourselves if we think that we know how the shipments are going to break down, shipment by shipment, over the next year, when we have a hard time predicting aggregate demand over the next month. And, if the customers are served by a different warehouse, then the shipment patterns may change as well. Running a model with many shipments rolled into a single demand point will be as accurate as one with individual shipments.

However, having said that, we do not have to blindly roll up every shipment into an aggregate number. We will discuss this in more detail in Chapter 12 and Chapter 13 "Data Aggregation in Network Design." But, for now, you can think about refining your demand points to categories like the following:

- Total demand to this point arriving in full trucks
- Total demand to this point arriving in trucks that are half full
- Total demand to this point arriving via overnight express

There are techniques for breaking down the total demand into smaller buckets, but keep in mind that your goal is accuracy and not false precision.

Transportation Costs Per Unit

As we pointed out, the underlying mathematical model has transporation costs expressed in terms of a cost per unit. If you look at any standard transportation reports or get quotes from carriers on rates, you will not see rates expressed as a cost per unit. It will be a cost per load, cost for the pallet, or cost per package to name a few. If we expressed the cost in terms of cost per shipment, we would fall into the same problem mentioned above—we would have to model every single shipment (and come up with a cost for every different shipment).

Luckily, with a little math, we can easily convert standard transportation rate structures into a simple cost per unit.

First, for perspective, let's discuss the many different transportation cost structures. These structures are driven by different modes of transportation (air, road, rail) and even by ways transportation companies quote their rates. When converting transportation rate structures to a cost per unit, we must understand the different types of rates and then apply them appropriately within our model. In practice, commercial network design software will convert a standard rate structure to a cost per unit for the model. However, as a modeler, you must have a good understanding of the rate structures to ensure that the costs used by the model accurately represent reality.

For the purposes of the model, we are interested in the cost to go from facility i to customer j. However, many different transportation providers may make multiple stops between picking up product at facility i and delivering it to customer j. For example, if I order something from a large online retailer that is shipped via UPS from the retailer's warehouse to my house, UPS may route the product through several of their own warehouses and hubs before actually dropping it off at my doorstep. But, for the sake of modeling, the retailer is only concerned about the single rate that UPS charges them for the delivery and not all the internal UPS movements that take place to ensure that delivery. Here are some of the different modes of transportation and how the rates are typically provided:

- **Commercial Truckload or Full Truckload (TL or FTL)**—In this mode, a firm is hiring an entire truck to drive from one location to another. The firm is charged for the use of the full truck whether the truck is 100% full or 25% full. The firm does not have to worry about where the truck comes from before the delivery or where it goes after. The trucking company arranges that as part of their own operations. These rates are typically structured either as a flat rate from one point to another or as as a cost per mile or kilometer. If the carrier charges a cost per distance, they may sometimes add a fixed charge on top of that and enforce a minimum charge to prevent very low costs for short moves. The rates charged can vary depending on where the shipment starts and ends. So, two shipments with exactly the same distance, but with different starting points, may have different rates. The carrier contract may also include a separate fuel surcharge which allows the trucking company to manage fluctuating fuel prices without having to recreate each firm's rates each time. Each of these trucks will have a capacity typically measured in total weight, total cube, or total number of pallets. Also, the term Truckload often lumps a lot of truck types together under the same umbrella. When you are building your model, you may want to consider different types distinctly—regular dry van, refrigerated, flatbed, tanker, 48- and 53-foot trailers, as well as other distinctions that affect the rate applied or the amount of product that can fit within each. Transit time for truckload shipments is normally easily predicted and dependable as less touches are required during transit and it's a straight shot from one location to the other.

- **Private/Dedicated**—A private fleet is one in which a firm owns its own trucks. These trucks are used to either deliver full loads to the customer or make multiple stops. A dedicated fleet is similar to a private fleet except that it is owned (and maybe operated) by a third party but dedicated to a single client. The rates for private or dedicated are usually a cost per mile with an optional fixed cost (similar to commercial TL transport described above). Because the firm owns the trucks, they do not need to worry about the rate changes based on the direction of the moves as they would with the use of TL transport. But, they do need to worry about having the truck return. Often, with a private fleet, firms can use the trucks to make deliveries to their customers and then also pick up product from their nearby suppliers on their way back to the origin (this is commonly known as a roundtrip or backhaul). Transit times for this mode are similar to Commercial TL except for the fact that there may be a delay in shipments waiting for trucks to return. If a firm has issues with not having enough vehicles available when needed, they will often supplement their Private/Dedicated transport with Commercial TL transport as well.

- **Less-than-Truckload (LTL)**—The LTL mode refers to a different transportation market. In the TL market, a truck comes to a firm's dock door, picks up the product, and directly delivers it to the destination. That truck is dedicated to that firm's load only. Contrary to what might make sense based on the names, if the firm simply does not fill up the entire truck, it is *not* referred to as an LTL move. It is still a TL move. A firm goes to the LTL market when it has a small load to ship to a destination where it is not economical to hire an entire truck to make the delivery. In this market, the LTL trucking company picks up the firm's small load on a route with other firms' pickups and deliveries. This load then rides with other loads through multiple hubs, where it is placed on different trucks with loads having similar geographic destinations, until eventually loads are placed on a truck departing a final hub to make the trip to complete a route of deliveries to customer locations. However, a firm does not need to worry about these intermediate activities completed by the LTL carrier. The LTL trucking company quotes a single price from origin to destination even if they route the product through different hubs each time. In the U.S. and Canada, these rates are usually based on the source, destination, weight of the shipment, and shipment class (a measure to define the density of the product, whether it is hazardous, and other measures that impact the cost for the carrier to move the item). In Europe however, the LTL market is typically quoted by the number of pallets and the distance it must travel from the source. There are break points in the rates based on some form of capacity level (the more product shipped, the lower the cost per unit to deliver it). Also similarly to TL rate structures, minimums and fuel surcharges are often assessed. As a firm ships larger and larger capacities per shipment a break-even point between shipping via LTL and shipping via TL can be calculated. For example, a TL carrier may hold 45,000 pounds of product, and shipping anything under 15,000 pounds turns out to be cheaper when using the LTL transport mode and rates. LTL shipments typically experience longer transit times however. For the economics to work, the LTL firm will have to send your product through a series of hubs where it may need to wait for other shipments going in the same direction as your load before its final delivery takes place. LTL firms make money when they can fill up trucks with many small shipments and therefore time to destination is sometimes sacrificed in lieu of filling up a truck with more loads per route.

- **Parcel**—The parcel mode is a lot like LTL but with even smaller shipments sizes. Here a firm is typically charged based on the source, destination and package weight. For heavier packages or shipments, this can be quite an expensive transport mode. However, for packages up to about 150 pounds, it can actually be much cheaper than shipping via LTL. This is because LTL firms are not set up to handle small shipments like specialized parcel carriers. With parcel, there are many service options that give you control over the shipping time, However in most cases, the faster the service the higher the corresponding price. If you want it overnight, the parcel services leverage their more expensive air networks to fly your package. If you are willing to wait, the package moves through their more economical ground network.

- **Ocean**—Ocean transport is similar to truckload transport but a firm's product moves in containers (as opposed to truck trailers) across the ocean between port cities (opposed to along highways and local roads). Ocean rates are usually quoted as the cost of a single container from one location to another. The costs can be from port to port or from door to door (where the truck or rail transport to the first port and from the last port is included in the rate). The containers can also vary in size (40-foot, 20-foot) or in capability (refrigerated or ambient). Ocean transit time can be quite long because ocean vessels follow fixed schedules making stops at many different ports along the way.

- **Rail**—Rail transport moves a firm's product from one railhead (railcar loading/unloading location) to another. Rail rates are typically expressed in terms of the cost of a rail car to get from one point to another. Like ocean transport, these rates can be between two railheads or from door to door. Most frequent rail shippers have rail lines directly to their plant or warehouse locations. From a transit-time perspective, rail shipments take longer than over-the-road moves (TL, LTL) due to the number of switches and connections required along the way. As a result, they are typically advantageous for cross-country moves with products that are not time-sensitive.

- **Intermodal**—Intermodal transport combines TL and rail transit. An intermodal move involves a firm's shipment originally being loaded onto a truck trailer and then delivered to a railhead for transfer onto a flatbed rail car for a portion of the journey. The transfer back to a truck at the destination railhead for final delivery may also take place if the final shipment destination does not have access to its own railhead. These rates typically are expressed as door-to-door rates because the shipping firm handles the full move from origin to final destination. This type of transport can be very economical for cross-continent moves. In terms of costs, intermodal typically falls in between rail and TL transport costs—that is, intermodal is cheaper than TL but more expensive compared to rail. From a transit-time perspective, intermodal takes longer than TL but is faster than rail.

- **Multistop**—The multistop mode is like the truckload moves, but involves several stops. The multistop mode cost falls between TL and LTL above. In this mode, a firm has several shipments to deliver. The shipments are not all large enough to use a full TL for each however. But together they are too big for an

LTL shipment as well. In multistop transit a truck picks up a firm's load at one facility and then makes deliveries to several destinations. Unlike LTL, the truck is dedicated to just this firm's loads. The rates are usually expressed in terms of a cost per mile or kilometer with a stop-off charge. The stop-off charges may be waived for the first stop and commonly increase with the number of stops being made on a single route. Often, a trucking company will actually put a limit on the number of stops a single truck will make. There also may be restrictions on the "out of route" miles a single truck may travel within a single route. Out of route is a measure of how far the route deviates from a direct route from the source to the final destination. Multistop transit times are obviously longer than TL which make no intermediate stops from origin to destination. However, multistop transit is typically faster than the use of LTL as the firm has more control of the routes being made and the use of hub locations is much less common. We will discuss multistop in more detail later because they do not always allow for a straight forward calculation of the cost to go from one point to another.

As a side note, a frequent question that arises is: Why not just use the cheapest transportation mode? In general, the more you ship at one time, the lower your transportation costs. That is, if you ship the same total amount, the LTL cost will be lower than parcel, the TL lower than LTL, and Rail lower than TL. In fact, there can sometimes be a nice savings from shifting modes. However, when you ship a quantity too small to be economical for a certain mode, your costs will actually skyrocket. For example, if you hire a full truck to ship a single small package, your cost for that single package will turn out to be quite high. In other cases, you have to make a trade-off between transportation costs and inventory. For example, if a retailer receives LTL shipments once a week from its suppliers, shifting to full truck transit means they shift to only receiving one shipment every six weeks. This change decreases the transportation costs, but will increase the inventory. It is beyond the scope of this book, but firms do mode selection studies frequently to understand the best modes and optimize the trade-off between transportation and inventory.

Now that we understand the different modes and their rate structures, we need to convert the costs to a cost per unit. You should also see that converting to a cost per unit in a mathematical program allows us to standardize the input data. Although some rate structures may require an extra step or two, the basic formula for the cost-per-unit is shown as follows:

$$trans_{i,j} = \frac{load_{i,j}}{avg_{i,j}}$$

Here $trans_{i,j}$ is the unit transportation cost we need for our optimization problem, $load_{i,j}$ is the cost for the load to move from point i to j, and $avg_{i,j}$ is the average size of the load. As you saw from the above list of modes, most cost structures will directly give you the $load_{i,j}$ cost. And, if not directly, within a simple step or two.

For some shipment types, determining the $avg_{i,j}$ is trivial. Assume our units of measure are in pounds. For parcel and LTL, the cost per load is dependent on the size of the shipment. For a ten-pound parcel, the cost of the load will be based on the fact that it is a ten pound shipment and the average shipping size is ten pounds. For an 8,000 pound LTL shipment, the cost of the load is based on an 8,000 pound shipment and the average shipment size is 8,000 pounds.

For TL (and rail, intermodal, ocean, private/dedicated—which are calculated like TL), the average size is the amount that is actually shipped. So, in these modes you are charged to move the whole truck or container or rail car. So you need to figure out how much you actually shipped on the load. For example, you may be given a cost of $800 for a load to go from point A to B. The TL can hold 45,000 pounds. If you ship, on average, 25,000 pounds, that will be the value you use for $avg_{i,j}$.

Let's walk through some examples of how this may be done.

In the example shown in Figure 6.1, we have shown how to calculate the TL load cost based on the shipment distance, the cost per mile, and associated minimum charge. The load cost is based on the larger of the distance rated cost and the minimum charge. For the Chicago to Dallas lane, the distance rated cost ($1.80/mile * 967 miles) was much larger than the minimum charge, while the cost for the Chicago to Indianapolis lane was based on the minimum charge because this was larger than the distance rated cost.

Shipment Origin	Shipment Destination	Shipment Wt (Lbs)	Distance (Miles)	TL $/Mile	Min Charge	Distance Rated Cost	TL Load Cost (max of rated cost and min charge)	Trans Cost per Lb
Chicago, IL	Dallas, TX	38,000	967	$1.80	$400	$1,741	$1,741	$0.046
Chicago, IL	Indianapolis, IN	27,000	183	$1.80	$400	$329	$400	$0.015
Atlanta, GA	Miami, FL	40,000	663	$2.00	$375	$1,326	$1,326	$0.033
Atlanta, GA	Chattanooga, TN	30,000	118	$2.00	$375	$236	$375	$0.013

Figure 6.1   Example for Calculating Transportation Cost Per Unit

The determination of the average shipment size ($avg_{i,j}$) requires more careful thought however. Of course, we want to determine the average shipping cost over all demand. That is, the cost, $trans_{i,j}$, is the cost to go from point i to point j to satisfy all the demand at point j that is satisfied from shipments from point i. It is not just the cost for a single load. So we need an accurate measure of the average shipment size ($avg_{i,j}$). For example, in Figure 6.2, we see a sample of lanes and the associated total weight that was shipped along them. By determining the average weight loaded into the vehicles during each shipment, we have an accurate measure of shipment size.

Shipment Origin	Shipment Destination	Total Truckloads	Total Weight Shipped (Lbs)	Avg. TL Weight (Lbs)
Tulsa, OK	Houston, TX	26	832,858	32,033
Tulsa, OK	Albuqurque, NM	35	1,246,175	35,605
Tulsa, OK	St. Louis, MO	54	2,084,508	38,602

Figure 6.2 Sample Lanes and Average Truckload Shipment Size Data

A commercial network design package will help you with many of these steps. However, it will not do all the thinking for you. You will have to make decisions on the best shipping sizes to use and how to translate your raw data into a working model.

We should keep in mind that we do not want to over-engineer this solution. We want an accurate estimate of the costs, not a precise estimate. The profile of shipment sizes from last year is not likely to exactly match the profile of shipment sizes next year. So close enough is good enough.

Determining All the Transportation Costs

Our third note about the transportation costs was that we needed a rate from every source to every destination. The challenge is that you may not have access to all the transportation rates you need for the model. This could be for a variety of reasons:

- You want the model to evaluate lanes that you are not currently using. As a starting point, you have data on the lanes you are using, but not on the new lanes you want to evaluate.

- Even if your carriers provide you rates, they may only give out "retail" (or non-negotiated) rates until you agree to use that lane and then they will work with you to arrive at a negotiated rate based on the total volume of freight on that lane. The problem is that your existing rates on existing lanes represent negotiated rates. Presumably, if you use new lanes, you would end up with negotiated rates on those lanes as well.

- Your internal data sources are not clean enough to use. The way your data is stored, it is not reliable to determine rates or average shipment sizes.

- You do not have time to gather the data.

In these situations, you may need to estimate the shipping costs. You can do this in various ways:

- Use a simple cost per mile or per kilometer based on the average rates you are paying today. For LTL and parcel, pick a single average shipping size and use the cost estimates tied to that number.

- Use nonnegotiated rates and extrapolate. Often, shipping companies will provide you with a matrix of standard or nonnegotiated transportation rates. Alternatively, there are online services where you can purchase this type of data for specific modes. Then, you can take this information and apply it to all your lanes (current and potential). These rates will most likely be higher than what you are actually paying today. But the trick is to then take the output lane costs for your current lanes and reduce (or increase) them by a uniform percentage so that the total cost on the lanes you use matches the total cost your historical data says you paid. Then apply this blanket percentage reduction (or increase) to all the rates in your purchased data for use within your final model scenarios. This can be an effective way to quickly get accurate rates.

- Use high-level industry benchmarks. We have had success in developing accurate rates using industry-wide average cost per ton-miles published by various independent sources. These rates are commonly expressed as the average cost to move one ton of an item one mile. We have found that generally, in the U.S., the cost per ton-mile in the range of .08 to .11 works well for TL and 0.31 to 0.35 works for LTL. Multistop rate estimates are usually somewhere between the TL and the LTL rates. Parcel rates estimates can be scaled to by three times the LTL rates, and conversely rail tends to only be 50% to 80% of the TL rates. These general rates are good, on average, but have some shortcomings for very short or very long moves.

- Use regression analysis. When you have a lot of shipment data to work with, you can use regression to determine good rates. A simple regression plots the distance of the shipment by its cost. So you can then develop a regression formula to determine a cost based on how far the shipment goes. See the example in the next section.

As an example of using industry benchmarks or published rates in the U.S., you will sometimes find rates provided to and from different states. And sometimes the state designations may be broken down even further. This is because there may be variations in $/mile rates even within a state (from a given origin). This is typically seen in relation to larger states with differences in economic activity across various parts of the state—for example, Northern Illinois (metropolitan Chicago area) has significantly more economic activity than Southern Illinois, thereby yielding a lower $/mile rate. In these cases, it is very common to split a state into subregions—say, California-North and California-South as an example. This provides a better and lower level of granularity. You will need to determine for yourself whether this granularity provides additional significant digits or is just a rounding error however. A sample transportation rate matrix is shown in Figure 6.3 with some states broken into smaller regional groupings.

Cost/mile	Destination						
Source	AL-N	AL-S	AR	AZ-N	AZ-S	CA-C	CA-N
AL-N	$ 1.99	$ 1.97	$ 1.72	$ 1.63	$ 1.63	$ 1.45	$ 1.47
AL-S	$ 2.04	$ 2.04	$ 1.74	$ 1.68	$ 1.68	$ 1.47	$ 1.51
AR	$ 1.68	$ 1.72	$ 2.06	$ 1.65	$ 1.65	$ 1.51	$ 1.53
AZ-N	$ 1.23	$ 1.23	$ 1.29	$ 1.76	$ 1.76	$ 1.27	$ 1.28
AZ-S	$ 1.23	$ 1.23	$ 1.29	$ 1.76	$ 1.76	$ 1.27	$ 1.28
CA-C	$ 1.25	$ 1.30	$ 1.28	$ 1.97	$ 1.97	$ 1.76	$ 1.89
CA-N	$ 1.20	$ 1.25	$ 1.23	$ 1.80	$ 1.80	$ 1.72	$ 1.74
CA-S	$ 1.28	$ 1.35	$ 1.25	$ 2.21	$ 2.21	$ 1.89	$ 1.97
CO-N	$ 1.16	$ 1.16	$ 1.16	$ 1.39	$ 1.39	$ 1.18	$ 1.25
CO-S	$ 1.16	$ 1.16	$ 1.16	$ 1.39	$ 1.39	$ 1.18	$ 1.25
CT	$ 1.06	$ 1.06	$ 1.08	$ 1.31	$ 1.31	$ 1.33	$ 1.34
DE	$ 1.12	$ 1.12	$ 1.12	$ 1.34	$ 1.34	$ 1.29	$ 1.34

Figure 6.3   Sample Transportation Rate Matrix

Let's also walk through an example of calculating an average rate for a sample set of data.

In Figure 6.4, we have calculated the cost per mile for each lane in our sample, and an overall average rate as well as a rate by state. To test whether the average is a good measure, we calculate the standard deviation and the coefficient of variation (Std Dev / Average) for both population samples. The standard deviation gives us an absolute measure of variability and the coefficient of variation normalizes the measures and expresses it as a percentage.

Shipment Number	Origin City	Origin State	Destination City	Destination State	Shipment Wt (Lbs)	Freight Cost	Distance (Miles)	$/Mile	Avg $/Mile	Std Dev $/Mile	Coeff. Of Variation
1	Atlanta	GA	Toledo	OH	39,800	$ 1,286	663	$ 1.94	$ 1.91	0.05	3%
2	Atlanta	GA	Columbus	OH	33,503	$ 1,043	567	$ 1.84			
3	Atlanta	GA	Lima	OH	38,138	$ 1,149	586	$ 1.96			
4	Atlanta	GA	Youngstown	OH	30,500	$ 1,376	740	$ 1.86			
5	Atlanta	GA	Canton	OH	34,313	$ 890	461	$ 1.93			
6	Atlanta	GA	Raleigh	NC	32,175	$ 625	393	$ 1.59	$ 1.68	0.08	5%
7	Atlanta	GA	Charlotte	NC	37,409	$ 417	245	$ 1.70			
8	Atlanta	GA	Wilmington	NC	34,305	$ 730	415	$ 1.76			
9	Atlanta	GA	Elizabeth City	NC	36,695	$ 996	576	$ 1.73			
10	Atlanta	GA	Rocky Mount	NC	39,253	$ 742	464	$ 1.60			
							Aggregate Statistic		$ 1.79	0.14	8%

Figure 6.4   Cost Per Mile Estimation Based on Average Across Lanes

With this analysis, we see that the overall average rate is $1.79. We would again need to expand our analysis to determine whether we need to worry about the fact that the rate is different to North Carolina versus Ohio.

Regression Analysis for Building a Rate Matrix

We will now go through a small example to help us understand how to analyze shipment data and estimate transportation rates using regression analysis. For this example, we have taken an extract from a shipment file, pulling sample shipments from Atlanta to Ohio and North Carolina. Regression can be valuable for helping determine a good cost to use for sources and destinations where you have no current shipment activity or rates.

The shipment table, shown in Figure 6.5, shows the origin location, destination location, shipment weight, and freight cost for each shipment. These are truckload shipments, as can be seen from the shipment weight. As mentioned previously, it is typical for shipments over 15,000 pounds to be delivered using truckload (versus LTL).

Shipment Number	Origin City	Origin State	Destination City	Destination State	Shipment Wt	Freight Cost
1	Atlanta	GA	Toledo	OH	39,800	$ 1,286
2	Atlanta	GA	Columbus	OH	33,503	$ 1,043
3	Atlanta	GA	Lima	OH	38,138	$ 1,149
4	Atlanta	GA	Youngstown	OH	30,500	$ 1,376
5	Atlanta	GA	Canton	OH	34,313	$ 890
6	Atlanta	GA	Raleigh	NC	32,175	$ 625
7	Atlanta	GA	Charlotte	NC	37,409	$ 417
8	Atlanta	GA	Wilmington	NC	34,305	$ 730
9	Atlanta	GA	Elizabeth City	NC	36,695	$ 996
10	Atlanta	GA	Rocky Mount	NC	39,253	$ 742

Figure 6.5 Shipment File Sample for Regression Analysis

Our objective is to use this shipment data to derive a $/mile rate that can be used within the model (as we were only given overall shipment costs within the data available).

As a first step, we will calculate the distance for each shipment (or lane). We will then plot the shipments on a chart with freight costs and distance, and perform a regression analysis (using capabilities within Excel).

We will perform the regression analysis (and plot the chart) using the entire sample of ten lanes. This is shown in Figure 6.6. The chart shows that the data fits well along a linear line based on the regression analysis. This shows a calculated rate of $1.83/mile. The regression yields an R^2 value of 0.95 and statistically significant variables (which you can see if you repeat this example in Excel and explore the output) which validates that this is a good fit.

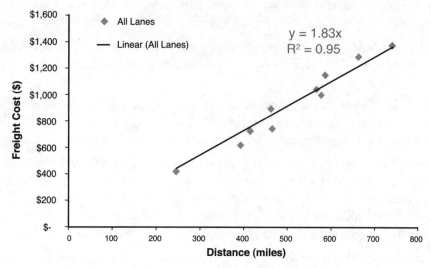

Figure 6.6   Regression Analysis Based on All Shipments

Here we have simply derived a cost-per-mile rate from the sample shipment data using regression analysis. Of course, in practice, we would want to do a more detailed statistical analysis. This can be accomplished by extending the same type of analysis across multiple origins and destinations to create a rate matrix, showing cost per mile rates on a more detailed state-to-state basis.

Estimating Multistop Costs

In many cases, the transportation costs are based on a point-to-point structure. This fits our model formulation quite well. In the model, we are assigning a customer point to a warehouse and calculating the cost to deliver each unit of demand. Using the methods already discussed, we can develop very accurate costs for this assignment. And this is true even when that customer is assigned to a different or new warehouse.

However, it becomes more complicated when calculating the costs for the multistop mode. Multistop is commonly used by retailers, but is applicable in other industries as well. With multistop, the final warehouse may send out a truck that makes several stops at different stores on the same route. So the transportation costs are no longer a simple point-to-point rate. The transportation rates will depend on which stores are on which route. And there is no one correct way to allocate the costs of the routes to each individual stop.

If the routes never changed, we would not have a problem because we could just pick a way to allocate the total route costs to each of the stores and we could easily produce accurate costs.

However, the whole point of a network design study is to consider new configurations for the supply chain. This means new warehouses, new assignments, and ultimately new routes.

You might think we would simply change our model formulation to build routes to the customers from the warehouse assignments. We agree that this would be a great approach. However, the limits of mathematics and computation complexity get in our way.

Remember, the number of potential combinations in a network design problem can explode quickly. In Chapter 3, we saw that if we want to pick the best 10 facilities out of 25, we have more than 3 million combinations. If we want to pick the best 10 facilities out of 250, we have 3 million *multiplied by* 32 billion combinations. It gets even worse very fast if we have to determine the best route out of each of these combinations as well.

Now, let us assume that a warehouse has 50 stores to service and that the routes have five stops each. Using the Excel formula PERMUT gives us the number of total permutations of picking out 5 stores to visit out of the 50 possible. We use PERMUT because any store could be on any route in any order. This yields a total of 254,251,200 possible routes for 5 stores. After that, we still have 45 more stores to cover. If we use the same formula for 45, 40, 35, and so on, we end up with more than 500 million routes to analyze. And this doesn't include the possibility for routes with 4 stores or 6 stores. If we had 100 stores instead of 50, the number of routes would be more than 34 billion. So a 2-times increase in the number of stores leads to a 68-times increase in the number of possible routes.

If we were to solve both of these problems together, we would effectively need to multiply the number of combinations of warehouse choices by the number of possible routes.

Besides the size issue, there are also differing levels of detail in the routing versus network design problems as well. In our network design problem, we are typically concerned with annual demand and long-term decisions. In routing, we are typically looking at a day's or week's worth of shipments. And these shipments may change from day to day or week to week. To make sure that the routes are realistic, we need to worry about things like the shipment sizes, time windows for deliveries, driver rest rules and many more operational constraints.

These problems are simply too large and too complex to combine into one model. And, more important, we do not have to combine them in order to solve our network design problems. We can still develop accurate transportation costs, get good answers, and determine insight into our multistop routes by following the basic principles of a robust multistop approach as follows:

- Worst case, the costs of multistop routes are somewhere between the cost of TL moves and those of LTL moves. For example, we previously mentioned that a reasonable estimate of TL rates is between $0.08 to $0.11 per ton mile and LTL between $0.31 and $0.35. So, multistop rates around $0.15 to $0.25 may be accurate for your model.

- If we compare the equivalent costs of the point-to-point rate with the cost of the route, we can simply adjust the costs with a scaling factor and produce a good answer.

- The network optimization problem is typically a much bigger impact decision and can safely be run first to develop a few alternatives. Then, with each of those alternatives, you can run separate and detailed routing studies to essentially break the tie. Occasionally though, these routing studies may lead you to adjust some costs in your network design model and rerun a few scenarios as well.

To illustrate our approach, we will start with a simple example. We will look at a multi-stop route from a warehouse to three stores. We will plot the points on a grid and assume straight-line travel distance. We will also assume that costs are directly proportional to distance.

In all the examples, the warehouse is at point (0,0). In all the examples, the warehouse is at point (0,0) and the trucks must all start and end here.

In the first example, shown in Figure 6.7, the three stores in order of the deliveries are at (6,6), (5,10), and (3,8). If we calculate the total distance of this route, we have to calculate the distance between (0,0) and (6,6), between (6,6) and (5,10), between (5,10) and (3,8), and between (3,8) and (0,0).

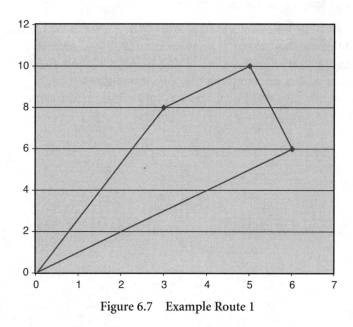

Figure 6.7   Example Route 1

If we use the simple straight-line distance formula $Dist = \sqrt{(x_1 - x_2)^2 + (y_1 - y_2)^2}$ to calculate the distance between two points for our example, (x_1, y_1) and (x_2, y_2) the total distance (and our estimate for cost) comes out to be 23.98.

In our network design formulation however, we only consider the distance from the warehouse directly to each of these points separately. So, how do we compare our total actual distance of 23.98 with how our mathematical model would treat this?

In the mathematical model we consider the distance out and back to these same points. That is, we use the same formula to calculate the distance to each customer, multiply that by two for the trip back, and then add the three distances together. This gives a total distance of 56.42. Now, to make a fair comparison of the multistop route to the 56.42, we need to assume that we run that route three times. That is, presumably we run a multistop route because we want more frequent deliveries. So, in this case, each route delivers one-third of the total requirements for each customer. Therefore, if we run the multistop route three times, we are making a fair comparison to the direct full truckload delivery distances of 56.42. Running the route three times gives us a total distance of 71.94 miles. So we could deliver each store their full requirement with a single truckload shipment, and that total distance would be 56.42. Or make three deliveries on multistop routes, for a total distance of 71.94. It makes sense that the route distance is longer for the multistop in this case—we get the benefit of more frequent deliveries at an additional cost.

In this case, if we want to approximate multistop costs using TL rates within the model, we simply multiply the TL rates by 1.28 (71.94/55.42). Then, our costs in the model will accurately reflect that we are running routes.

This factor will change based on the structure of the routes as well. If the stores are all far away from the warehouse, but close to each other, the factor may be much smaller. In the next example, if the stores are at (6,6), (6.1,5.9), and (6,5.8), then the total two-way distance is 50.6, and running the route three times is 51.3. The illustration in Figure 6.8 shows why these two distances are approximately the same. In this case, the factor is 1.01

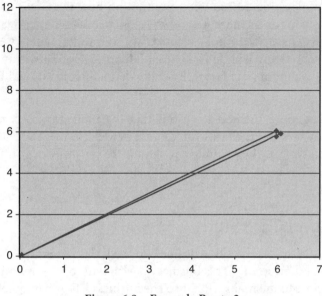

Figure 6.8   Example Route 2

If the stores are very far apart, as in the analysis shown in Figure 6.9, the factor will be 1.64.

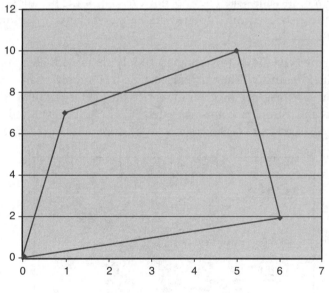

Figure 6.9   Example Route 3

These three examples provide you with intuition on how the factor changes based on the type of routes you run. If the stops are very close to each other and far away from the warehouse, you will have a very small factor. If you run routes with a very long distance between stops, your factor will be closer to 1.5. Experimenting with your data would help determine the appropriate factor. But note that a high factor will be between 1.25 and 1.75.

This section gives you a method for estimating the multistop costs to feed into the model. There may be other methods you can use as well. No method will be 100% accurate, but they will typically be accurate enough for the purposes of your network design work.

Transportation Case Study

This case study is based on an online business in the U.S. that ships products via UPS to customers around the country. This business started with one warehouse in Louisville, Kentucky, very close to the main UPS hub. The business has grown quickly and they are now considering opening new warehouses to offer better service to their customers. Right now, they ship about 3.3 million packages annually with an average shipment size of 2 pounds each. They originally thought they would ship more packages via UPS air, but so far their customers have paid for their own air shipments and therefore they will not include them in this analysis. They do, however, pay if the product ships via ground transport and the associated rates applied are based on the UPS Ground Commercial rates from 2008.

The map in Figure 6.10 shows their current supply chain with all shipments shipping out of Louisville, Kentucky. The current cost of this supply chain is $16.6 million. However, as you can imagine, the ground service they currently offer is a bit slow. The average distance to customers is approximately 975 miles. As expected, the breakdown by distance bands (see Figure 6.11) shows a lot of shipments traveling long distances.

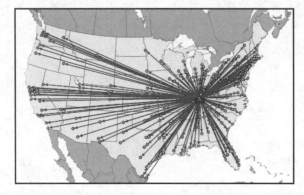

Figure 6.10   Transportation Case Study Baseline Network

Distance Band (Miles)	Total Packages	% of Total	Cumulative %
300	259,000	8%	8%
600	919,000	28%	36%
900	949,000	28%	65%
5,000	1,135,000	35%	100%

Figure 6.11   Transportation Case Study Baseline Distance Bands Summary

This firm wants to determine whether they can reduce costs by opening additional warehouses. They originally assume that adding two additional locations (for a total of three) would be about the right number. Although they are looking to improve their service, they will not make a move unless they can financially justify it first. They first run the optimization to find the best three warehouses.

As you can see in Figure 6.12, when the model is allowed to pick two additional sites, it selects Fresno, California, on the West Coast and Dover, Delaware, on the East. The total cost decreases to $15.2 million, for a savings of $1.4 million or 8% reduction in transportation costs in comparison to their current network. As you would expect, the distance to the customers improves as well. Looking more closely at Figure 6.13, we can see the details of this improvement. The average distance dropped to approximately 430 miles, a more than 50% reduction, and the customers within certain distance bands also improved.

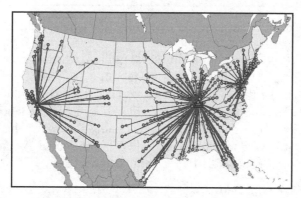

Figure 6.12   Transportation Case Study Best Three Warehouses Solution

Distance Band (Miles)	Total Packages		% of Total	
	Just Louisville	3 Distribution Centers	Just Louisville	3 Distribution Centers
300	259,000	1,347,000	8%	41%
600	919,000	1,088,000	28%	33%
900	949,000	476,000	29%	15%
5,000	1,135,000	351,000	35%	11%

Figure 6.13   Transportation Case Study Baseline Distance Bands Comparison

It is not surprising that the distance to customers improved. As mentioned previously in this chapter, transportation costs are strongly correlated with the distance traveled.

Like we've seen previously in this book, this firm now wants to understand the marginal value of a fourth warehouse. They determined that if they could get somewhere close to another $1.4 million in savings, it might be worth serious consideration.

When they run the optimization to consider the best four warehouses (results shown in Figure 6.14), the model adds Dallas, Texas, to the previous solution. Note that it did not have to select Fresno, California, and Dover, Delaware but in this case they were still optimal within this solution.

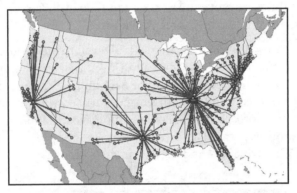

Figure 6.14   Transportation Case Study Best Four Warehouses

In this scenario, the costs decrease to about $15.0 million, which is only a $200,000 savings from their previous 3 warehouse solution (about a 1% reduction). This $200,000 difference also tells us the threshold for the fixed cost. That is, if the fixed cost of the additional facility is less than $200,000, it would still be cost beneficial to open the fourth facility in Dallas. This is a great example of being able to use the information on the fixed cost without having to actually include it in the model. As mentioned previously, we will discuss this concept further within Chapter 12.

In this case, the management team decides that it is worth it to explore adding two warehouses. The $1.4 million was a significant savings, and they could use this to build a strong business case to present to the board. The service improvements will also be a nice part of the business case, but the financial benefits of this are harder to quantify.

However, the management team also felt that they needed to further analyze the costs of the facilities. We will explore these costs in the next chapter.

Lessons Learned with Transportation

Transportation costs, like the weighted-average distance, tend to pull the location of facilities as close to the high demand customers as possible. This objective varies from solutions provided by minimizing weighted-average distance however when the transportation rates have large minimum charges or relatively flat rates (then, as long as you are within a radius of large points, you are okay) or when the rates are not symmetrical (then, it pulls the locations closer to the lower transportation rates as well as high demand).

Transportation costs are often the most important costs in a network design study, and they are the first ones you should consider adding to the model.

These costs come in various forms depending on how product is moved through the supply chain. Although there are some special cases to consider, such as multistop routes and the use of regression to fill in missing rates, transportation costs come down to the fact that you need to apply the appropriate cost per unit to go from Point A to Point B.

It is good to keep in mind the lessons we learned about significant digits when building in transportation rates as well. The details of transportation can get quite complex. But, in the end, you cannot predict the cost of oil or exactly how your customers will order, so your measurements need not be precise enough to discuss costs down to the last dollar either.

When running and analyzing models with costs, keep in mind that the lessons we learned from the models with just distance still apply. It is also good to run multiple scenarios; the additional scenarios can give you the value of an additional facility, and the service metrics still apply.

Finally, we can't forget that adding the costs does usually help you create a better business case as well.

End-of-Chapter Questions

1. UPS Model—Mini Case Study

   We will work with the same case mentioned previously in this chapter, except average shipping size per package will change to five pounds. You can find this case and further instructions on the book Web site. The file is called UPS Model 5 lb Avg.zip. Open the model and change the average shipping size to five pounds.

   a. Run the baseline model with just the Louisville warehouse and keep the current demand. What is the cost of the baseline?

---

b. Run additional scenarios for the best two, three, and four warehouses. Which warehouses did it pick, and what is the cost difference between the scenarios?

2. TL and LTL Model—Mini Case Study

A distributor of all types of residential construction products (wood, nails, fixtures, appliances, windows, and so on) delivers product directly to the job site. For small jobs, like a single house, they ship in LTL quantities. For large jobs, like an apartment complex, they ship in TL quantities. They sell to customers across the U.S. and use an average TL rate of $0.11 per ton-mile (with a minimum charge of $12.50 per ton) and $0.30 per ton-mile for LTL (with a minimum charge of $4.00 per ton). The model, more details on the case, and directions on how to use it can be found in the file TL and LTL Model for Construction Products.zip on the book Web site.

a. In the existing supply chain, they are shipping everything from warehouses in Seattle, Dallas, and Pittsburgh. What is the current cost of this supply chain?

b. If they could pick any warehouses from the list of potential warehouses, what are the best three and what is the cost of the best three? What is the average distance to customers, what is the percentage within 300 miles, and how did the average distance to customer change?

c. If they were to add two more warehouses to the existing three, where should they put the warehouses? What is the cost of this solution? What is the average distance to customers, what is the percentage within 300 miles, and how did the average distance to customer change?

d. If they were to pick the best five warehouses, where should they put the warehouses? What is the cost of this solution? What is the average distance to customers, what is the percentage within 300 miles, and how did the average distance to customer change?

3. Transportation Costs and Oil Price—Mini Case Study

The fluctuating price of oil has a big impact on the cost of a supply chain. As oil prices increase, so does the cost of transportation. Recent fluctuations in the price of oil have taught supply chain managers that they need to develop a robust supply chain strategy that performs well if the price of oil is high or low. Also, supply chain managers need to be able to project the amount they will spend on transportation as the price of oil changes. This is a question that gets the attention of the senior management within any firm.

For businesses that ship mostly by truck, we can start to answer this question by understanding how the price of oil impacts the price of diesel fuel. (After we know this, we can approximate how much of our costs are driven by the price of diesel. For other modes of transportation, we can follow a similar methodology.)

Figure 6.15 shows a scatter plot of the price of diesel versus oil. This data can be found on the book Web site in the file `Oil and Diesel Prices.xls`. This represents weekly data from December 30, 2005, to March 30, 2012.

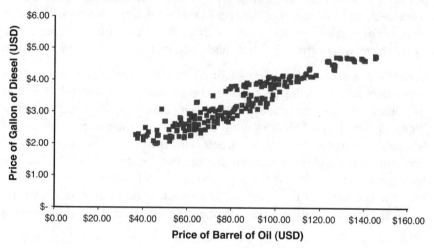

Figure 6.15   Diesel Versus Oil Cost Regression Analysis

A regression analysis can help us determine the relationship.

a. Run a regression in Excel on this data with the price of diesel as the dependent variable. What is the regression equation that relates the price of diesel to the price of oil? What is the R-squared value? What is the p-value for the independent variable, the price of oil?

b. How much does this model predict that diesel will increase for every $10 increase in the price of oil?

c. Build a chart that shows the expected price of diesel for every $10 increment in the price of oil from $20 a barrel to $200 a barrel.

d. If oil is currently $100 a barrel and you expect it to increase by 40% to $140, what percent increase would you expect in the price of diesel fuel?

e. If oil is currently $40 a barrel and you expect it to increase by 100% to $80, what percent increase would you expect in the price of diesel fuel?

f. How would you use this information when running different network modeling scenarios?

4. Consider a problem in which you need to be able to deliver emergency pharmaceutical products to the drugstores in Chicago. That is, if a drugstore runs out of an item and they need to fill an order, your service can immediately drive the product to their location. You want to determine where your facility should be located to service these drugstores. You also want to minimize the total transportation cost of providing this service. If your transportation provider has come back and told you that the cost to make a delivery is $40, why will this not help you determine the best location for your facility? If the $40 per delivery is all you have, what other objective might you consider to locate your facility?

5. Consider a consumer packaged goods company that sells their products to supermarkets. For the bigger supermarket retailers, they ship full truckloads from their warehouses directly to the retailer's warehouses. For smaller ones, they will ship their product in full truckloads from their warehouse to local wholesalers who then ship to the stores. This firm is interested in the best locations for their warehouses to minimize the transportation cost. If they have built a cost matrix that has the cost per truckload from every potential warehouse to every one of their ship-to locations, why don't they need to worry about the distance between the potential warehouses and the ship-to points?

6. When we build a model using expected demand for the next three years, we clearly realize that the demand will be a forecast. Why should you also consider the transportation cost a forecast? What does this tell you about the number of significant digits of accuracy you need for transportation costs in your model?

7. If the cost of an ocean container from China to Seattle is $3,500 and you can fit 250 units into the container, but on average you put in 200, what is the cost per unit that should be used within a network design model?

8. Go to the spreadsheet Raw Transportation Rates.xls on the book Web site. This file shows 300 raw shipments from the warehouse in Atlanta to customers located in Chicago. It shows a mix of TL, LTL, and parcel shipments. Determine the average cost per pound for shipping to Atlanta for each of these modes.

9. If you are working for a firm that ships with multistop truckloads, why might a cost of $0.22 per ton-mile be a good estimate? Assuming that it is a good estimate, why would you want to pick a slightly higher number, or a slightly lower number?

10. In the formulation of this optimization problem, we put in a constraint that limits the number of warehouses to P. If we did not have this in the model, why, in general, would the optimization choose to open and use every potential warehouse? Why is it likely to be a bad decision to open and use every potential warehouse (even though this minimizes the transportation costs to the customers)?

11. In the formulation of the optimization problem, we are assuming that a customer receives many shipments per year. Assume that we are working on a model for a company that makes 25,000 shipments per year. If we model each shipment as a customer point in the model, would this make the model more accurate? (*Hint: think about the ability to forecast future demand at this level.*) Would this make the model easier to work with and easier to analyze?

12. A manufacturer and distributor of flooring products including carpet, tiles, and hardwood uses a third party with a dedicated fleet to deliver to its customers. The products are picked up from a local warehouse and deliveries are made to customers located within a specific region around the warehouse.

    The negotiated transportation rates are based on the following structure:

Distance	Min Charge	Cost per Mile
0–100 miles	$200	–
100–300 miles	$200	$1.50
300–500 miles	$200	$1.80
>500 miles	$200	$2.20

    The truck typically makes multiple deliveries on a trip, so the cost per mile is applied on the total miles driven.

    Does this rate structure make sense? Why does the rate increase with an increase in distance? For a customer located 350 miles from the warehouse, what is the transportation cost per unit (assume that there were 300 units per shipment for this customer and that this is the only customer delivered on this trip).

    If this company were to optimize the warehouse network using this rate structure in place, how would this impact the locations selected? Would the results be different if there was no minimum charge in the rate structure?

7

INTRODUCING FACILITY FIXED AND VARIABLE COSTS

This chapter discusses the costs of facilities. We will model these costs in two categories: fixed and variable. They can be applied to plants and warehouses.

The fixed costs are a one-time charge independent of the volume. This could be the cost of building the site; expanding the site; adding equipment to the site like extra lines or equipment in plants, additional racking or automation in warehouses; paying taxes; or staffing the location. The fixed cost could be thought of as a one-time capital investment or as an annual fixed operating cost. Capital investments are treated differently than fixed operating costs because they are depreciated over time and you do not necessarily need to justify the investment over a single year. Fixed operating costs represent costs that are incurred during the normal operation of the plant, line, or warehouse. In essence, these are costs that are incurred each year (or month) independent of how much volume is handled at the facility. We will discuss this topic later in the chapter.

The variable costs are those that depend on the actual number of units that are made at a facility or that pass through a facility. That is, to determine the total variable cost, you simply multiply the total units that are made at a facility or that pass through a facility by the variable cost. For example, if the variable cost of production is $2, then every unit adds $2 of cost. Note that different products may have different characteristics, for example different sizes, weights, or manufacturing requirements. In this case, you may use a simple conversion table to convert units to another measure before applying the variable cost.

In practice, there is no correct answer on what cost is "fixed" and what cost is "variable." One of the authors of this book was told by two vice presidents of supply chain from two different firms in the same week the following:

1.   "There is no such thing as fixed costs. All my costs are variable over time. I can change anything."

2. "There is no such thing as a variable cost. All my costs are fixed. After I open a warehouse, I have to fully staff it and cannot adjust that in any way that impacts my current budget."

Both views are valid and you can take either view when setting up your supply chain model. It is important to understand how these costs work to determine how you want to use them in your supply chain. Also, remember that we are not necessarily interested in the accounting definitions, but, instead, we want to use these costs in the way that makes the most sense for the types of decisions we want to get out of the model.

Let's look at the two extreme cases to understand how the costs work:

If you model all the costs at a facility as fixed, then the optimization has the incentive to minimize the number of facilities it opens (to minimize the total fixed cost) and, when a facility is open, it will focus on maximizing the amount of product put through that facility (which is why you may want to include capacity constraints).

If you model all the costs at a facility as variable, then the optimization has an incentive to use the facilities with the lowest variable cost (as long as extra transportation costs don't outweigh it) and no incentive to avoid adding additional facilities (so the model may choose to use many different facilities unless you limit the number of facilities with a constraint).

You should note that there are some advanced varieties of the fixed and variable costs that allow you to capture other details within the supply chain.

First, you can think about the fixed costs as a step function. That is, instead of just a single fixed cost, you may have several fixed costs you need to add, depending on how much product flows through or is made at the facility. For example, if you open a plant with one shift, you may have a fixed cost of $5 million. If you add a second shift, your fixed costs go up to $6 million but it gives you additional capacity. And if you add a third shift, costs go up to $7 million with the added capacity. This seems like a combination of a fixed and a variable cost, but in reality, it behaves like a fixed cost. The model will open it up for a particular size, and after it does, all things being equal, it has an incentive to then use as much of the available capacity as possible.

Second, you can think about the variable costs depending on the size of the facility, or the economies of scale. That is, as a facility gets larger, the cost per unit may go down. You capture this again with the step sizes and simply have a lower variable cost for each step size. For example, if you open up a warehouse with the capability to process 200,000 pallets a year, your cost may be $5 per pallet. If the facility can process 400,000 pallets per year, your variable costs may drop to $4.50 per pallet. Often, to prevent the model from opening the 400,000-pallet facility and using it for only 100,000 pallets' worth of product, you either model a fixed cost in addition to the variable cost or enforce a minimum flow through the facility. That is, you stipulate that you would have to send at least 200,001 pallets through the larger facility before you would even consider using it.

In the math formulation section that follows, you'll see how the fixed and variable costs balance against each other. And you will see how the use of options can allow you to capture the step functions in the fixed and variable costs.

Mathematical Formulation

Let's start with the mathematical formulation of the problem. Here is the new mathematical formulation of the problem with facility variable costs:

Minimize

$$\sum_{i\in I}\sum_{j\in J}(trans_{i,j} + facVar_i)d_jY_{i,j} + \sum_{i\in I}\sum_{w\in W}facFix_{i,w}X_{i,w}$$

Subject to:

$$(1)\sum_{i\in I}Y_{i,j} = 1; \forall j\in J$$

$$(2)\sum_{i\in I}\sum_{w\in W}X_{i,w} \geq P_{\min}$$

$$(3)\sum_{i\in I}\sum_{w\in W}X_{i,w} \leq P_{\max}$$

$$(4)\sum_{w\in W}X_{i,w} \leq 1\forall i\in I$$

$$(5)\sum_{j\in J}vol_{i,j}Y_{i,j} \leq \sum_{w\in W}cap_{i,w}X_{i,w}; \forall i\in I$$

$$(6)Y_{i,j} \leq \sum_{w\in W}X_{i,w}; \forall i\in I, \forall j\in J$$

$$(7)Y_{i,j} \in \{0,1\}; \forall i\in I, \forall j\in J$$

$$(8)X_{i,w} \in \{0,1\}; \forall i\in I, \forall w\in W$$

This model is quite similar to those we have discussed previously. We have added a few new concepts:

- The set W represents the set of facility options. Each option represents a different sizing decision for the same physical location. That is to say, if you have the choice between building a small, medium, or large warehouse at each potential site, then W would contain three entries. This set of options now allows us to model changes in fixed and variable costs for each option. To keep the formulation simple, we are showing only the change in fixed costs with the options, but the model can be extended to include the change in variable costs.

- The decision on whether to open a facility has been generalized to include the choice of facility option. That is, variable $X_{i,w}$ will be set to 1 if and only if facility i is opened with option w. Note that constraint (4) ensures that you don't select more than one facility option for each facility that is opened. For example, constraint (4) forces you to pick the small, medium, or large warehouse option, not all three or any two.

- We have added facility fixed (*facFix*) and variable costs (*facVar*) to the objective function. Now these costs are added to the transportation costs. So the optimization will consider all the costs. The optimization will want to incur a fixed cost only if it absolutely needs the capacity or if the savings from the variable or transportation costs make it worthwhile.

- We have also added limits to the capacity of a facility. This is captured in constraint (5). If you have fixed costs and no capacity constraints, the model may be tempted to open as few facilities as possible and send unrealistically high volumes through these facilities.

- Note that our model now allows the solver engine to determine how many facilities to open. Constraints (2) and (3) restrict the total number of opened facilities to be between P_{min} and P_{max}. Without facility fixed costs, there would have been no disincentive for our model to simply open all facilities.

Facility Variable Costs

To understand how variable costs impact the model, it is a good thought exercise to think about what will happen if they are the same at every facility. If the variable costs are the same at every facility, the models we run will pick the exact same set of facilities and the exact same flow as the models we ran with just transportation costs and capacity constraints. Why is this? By adding a variable cost that is the same everywhere, the model has no incentive to pick one location over another. In fact, in the previous models, the variables were, in fact, the same, zero. The total cost of our solution has gone up because of the variable costs, but our decisions are the same. There is little use adding the same variable cost at all sites unless it is really no trouble.

Let's continue our case from the preceding chapter by adding variable costs to the model. In this case, the firm estimates that they pay $2.75 per package that they handle in the Louisville facility. If we simply add this to the baseline, we add $9.0 million in variable warehouse costs to the $16.6 million in transportation costs. The total cost is now $25.6 million. Because the Louisville facility is older, the firm believes that the variable costs will drop to $2.50 per package in a new location.

When we run the model, the solution is not the same as the run without the variable costs. The three locations are the same (Louisville, Kentucky; Fresno, California; and Dover, Delaware), but the territories have shifted. When you look at the map shown in Figure 7.1, you see that Dover's territory has expanded (compare back to Figure 6.12). It has picked up more territory on the East Coast and Florida and some regions around Wisconsin and Texas.

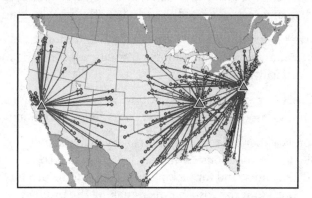

Figure 7.1   Revised Outbound Distribution Solution with Variable Costs Added

Why has this happened?

The objective function considers both the transportation costs and the variable costs. It does not consider distance to the customer. Also remember from Chapter 6, "Adding Outbound Transportation to the Model," that transportation rates may be relatively flat and not directly correlated with distance. And, with small-parcel rates, you may have a lot of rates that are exactly the same even though the distance from the warehouse is different. Then, the lower variable cost at Dover will push the assignment to that location. If the variable cost difference were even higher, the Louisville facility would have an even smaller territory.

So when the variable cost is different between locations, the decision can change and it can impact your model. Now, the model is making a more nuanced trade-off. All things being equal, the model will move more volume to the low-cost warehouses. When the optimization decides to move more volume to these low-cost locations, transportation costs will rise (or, best case, stay the same), but the overall result is a lower cost.

Determining the variable costs in a model is not always straightforward.

In some cases, especially when you use third-party warehousing or manufacturing, you will have a simple variable cost to include. That is, you will know what the third-party firm charges you per unit produced or handled.

When the facilities are your own, however, you may need to consider the following elements when determining the variable costs:

- **Labor Costs**—This is the most obvious factor. If you don't model labor as a fixed cost, you need to determine the labor needed per unit. Firms may have an estimate of this. Some firms may use the very detailed Activity Based Costing methods to determine how much labor cost is incurred for each unit produced or handled. In any case, you need to include both direct and indirect labor.

- **Utility Costs**—Depending on the type of firm, and more true for manufacturing firms, the more product that moves through a facility, the more electricity and water the facility uses.

- **Material Costs**—In manufacturing, you will need to purchase additional material for each unit produced or consume packaging material in a warehouse.

Often, it is easier and cleaner to break out the costs by major activity to help quantify the costs. For example, in the warehouse, you may have separate costs for putting product away, for storing the product, and for picking and shipping the product to customers. In the plant, you may break out the costs between making the product and packing it.

When you have many different sites and there is complexity in breaking out the variable costs, a regression analysis can be used. In this case, the total cost of the site is the dependent variable, and the total throughput of the site is the independent variable. You can add other independent variables as the situation requires, such as product type or type of facility. But using throughput can provide helpful insight. When the regression runs, the slope of the line provides information on the variable cost. The y-intercept of the regression can provide insight into your overall fixed cost per facility.

As you can tell, there are many ways to break out the variable costs and each situation is unique. We do recommend that you start with a simple approach and then add sophistication as needed.

Facility Fixed Costs

The second facility cost type is the fixed cost of the facility. In previous chapters you've seen how you can get some fixed-cost information for free when you run multiple scenarios. And because it is a good practice to run multiple scenarios anyway, this is an analysis you get for free as well—so you might as well use it.

In summary, when you run a scenario with one more facility allowed than another scenario, the difference in cost is the fixed cost threshold. For example, if a model comes back with a four-facility solution at a cost of $10.8 million and the five facility solution is $9.1 million, then the cost of the extra warehouse needs to be less than $1.7 million.

This technique can be a great way to account for fixed costs in a model. This is especially true when making warehouse location decisions. A firm will have a good estimate on the cost of a warehouse and will just need to understand the threshold of adding facilities.

This technique is also powerful when a firm uses third-party warehouses. In these arrangements, the firm may pay only a variable cost to use the third-party facility. This does not mean that the true fixed cost is zero. It reminds us to think deeply about the fixed costs and how we will analyze our results. Each additional warehouse we add to the model increases the amount of management overhead, decreases our control over the supply chain, possibly increases the amount of inventory, potentially erodes our ability to replenish the warehouses with full truckloads, and may cause our suppliers to increase costs if they have to ship into multiple facilities. So it may be impossible or too expensive to estimate these fixed costs. But, again, the management team should be able to judge whether the savings from adding an additional warehouse likely outweigh these hidden costs. For example, if the transportation savings are five times higher than the entire management overhead costs, it is likely that hidden costs will not materially impact our savings.

Having covered the reasons for *not* including a fixed cost, there are reasons you may want to include fixed costs. These include the following:

- The fixed costs vary significantly from location to location.

- Capacity is an important consideration and you want to model multiple options. For example, you may staff a facility with one, two, or three shifts and the shifts are best treated as fixed costs.

- We need to address capital investment decisions to expand or improve existing locations or open new sites. When considering a capital investment as a fixed cost, you want to be careful to make sure you put the investment in the same time frame as other costs. So if you have an annual model (and therefore your transportation and variable costs are annual numbers), you want to annualize your capital investment so that there is a fair comparison of costs. Most firms follow accounting guidelines in allocating capital investment to a year.

The fixed-operating costs most often include items like the management of the facility, the cost to upkeep the building, the utilities (if not dependent on the number of units you flow through the facility), the number of shifts (if these are not modeled as variable costs), and the taxes on the building and land.

The fixed capital costs include investment costs like building the facility, adding lines, and adding equipment.

It is usually important to separate the fixed operating and fixed capital costs into two distinct buckets.

When you build your business case for implementing one of the results from your network modeling exercise, you often need to financially justify your decision with a net present value (NPV) calculation. In general, the NPV calculation treats the capital investment as a one-time investment at the start of the project. So it is treated separately from the other fixed costs. We have an example of this in the end-of-chapter questions.

Categorizing Fixed and Variable Costs by Analyzing Accounting Data

We have found that even with all the previous guidance, it can still be difficult to determine fixed and variable costs. This is because accounting systems are meant to report the financial data in a way that makes sense from a financial standpoint and not necessarily in the format appropriate for a supply chain engineer.

To make good decisions about the structure of our supply chain, we may need information on the fixed and variable costs. The accounting systems may just report on total costs. These same systems may attempt to break out fixed and variable costs, but each location in the supply chain may categorize things differently.

One method we have found successful is to look at a firm's detailed list of expense categories for a facility. This list may contain things like direct labor, overtime, maintenance labor, management, security, IT spend, electricity, and many other expense categories. These are the standard expenses that the accounting team tracks and reports on every month. From this list, the supply chain team, working with the finance, accounting, and other functional areas, goes through the list, item by item, and categorizes each cost into the percent fixed and percent variable. This then provides a standard way to use the accounting data for supply chain design work. For example, the electricity costs may be split 30/70 between fixed and variable. The idea is that some expenses are fixed (because you just have to keep the lights on) and some are variable (the more you process, the more electricity you need to run your machines).

As an example of this process, we worked with a client who had five plants making the same type of product. They had one plant that, according to their accounting systems, cost about 75% more than their other plants. They wanted to justify closing this plant. However, this company sold a very heavy product (and thus expensive to ship) and this plant was located very close to customers. After doing a detailed study using the process we described previously, we determined that this plant was actually very good at making one type of product and very bad at making the other type. When we ran the optimization, the results suggested not only keeping the plant, but actually increasing its production. In the current situation, the plant was making about 50% of each product type. Because the plant was very good at making one type of product and was close to the market, the optimization had that plant make much more of that product and hardly any of the product that it wasn't good at making. In the end, it made more total product at a much lower cost. The accounting system was failing them because it was reporting on averages. The analysis revealed the underlying cost structure and the optimization model was able to take advantage of that.

Lessons Learned from Adding Facility Variable and Fixed Costs

When adding fixed and variable costs to the facilities, the optimization now has an incentive to pick facilities with the lowest cost. When optimizing based on total costs, the optimization can pick sites with low fixed and variable costs if the facility cost savings are greater than the increase in transportation costs. So you can now get locations and maps that do not seem to make sense if you are expecting sites to be close to customers.

When you're modeling fixed and variable costs, it is important to think about them in the context of your model. The accounting definitions of these terms may not help you make the correct decisions.

For fixed costs, you need to be careful to separate the ongoing fixed costs (like keeping the lights on) from the one-time investments (like building a new building). For the latter, the one-time investments, you need to make sure you put the costs in the same units of time as the other cost elements in your model or you will not get realistic results.

End-of-Chapter Questions

1. You are working for a firm that makes a special type of fertilizer that is sold in bags. They sell millions of these bags a year. Because of the accounting system, they claim that their fixed costs go up $100 for every hundred bags. Would you model this as a fixed or variable cost in a model?

2. If it costs $10 million to build a new facility and you expect to depreciate the building over 20 years, why wouldn't you include the full $10 million when you are building a model with a year's worth of demand?

3. Open the file Warehouse Costs by Throughput.xls located on the book Web site. This file contains the total cost to operate different warehouses by throughput. Run a regression analysis and estimate the fixed and variable cost formula. Why might this approach work well for a firm with many warehouses and not very well for a firm with only a few warehouses?

4. A large CPG manufacturer initiated a network study to optimize its manufacturing network to determine where add a line to make a new product. The company had an existing plant in the U.S. it could expand and was considering new locations for the line in either in Mexico or Panama. The costs for the three scenarios are shown in Figure 7.2. The fixed cost for the new line is shown in millions of dollars. The cost of the product, the freight cost, and the duty costs are shown in terms of a cost per unit. The freight cost is the cost to get the product from the production source to the U.S. Distribution Center (DC).

Cost Category	US Plant	Mexico	Panama
Fixed Costs ($Millions)	$5.00	$3.00	$1.00
Production Cost Per Unit ($/Unit)	$3.00	$1.50	$0.75
Plant to US DC Freight Cost ($/Unit)	$3.00	$5.00	$10.00
Duty Cost ($/Unit)	$0.00	$0.00	$1.50

Figure 7.2   Data Used in Analysis

The total demand for the line product is 200,000 units.

When the model was run, the 200,000 units were made in Panama. The company was expecting production to stay in the U.S .because of the high transportation cost from Panama. Use the cost data provided to figure out why Panama was picked.

5. Open the Investment Decisions.zip located on the book Web site. The file contains more details on the model. This is a firm that has five plants around the country and is facing tremendous growth. They have an opportunity to invest in additional capacity and also replace older equipment with much faster equipment. In this case, the investment decisions can be thought of as a step-function of fixed costs. That is, they can add blocks of capacity at a time for a fixed cost.

   a. What year will they run out of capacity with their existing network?

   b. In Year 5, what is the optimal plan? How would you describe the solution?

   c. In Year 5, what other alternative scenarios should they consider?

6. You have completed a study and have found that you can save $7.5 million in the first year after closing two warehouses, opening a new plant, and serving your customers from different facilities. This study was done with one year's worth of data. The $7.5 million does not include the initial investment to close the warehouses and open the new plant. The initial investment is estimated to

be $20 million. For the company you work for, you must include the net present value calculation when you present the business case. For the NPV calculations, your firm uses a 15% interest rate and goes out only five years (the thinking is that too much changes in five years, and they prefer shorter paybacks).

a. Calculate the NPV using the preceding data. Assume the same $7.5 million savings per year. Based on the NPV results, is this a good project?

b. The firm is growing fast. They would assume that the savings grow 5% per year. That is, if they implement this solution, the savings in Year 2 will be 5% higher than the $7.5 million because they will be shipping much more product two years out. What is the new NPV?

c. Being conservative and going back to the original assumptions, how large would the initial investment have to be for the NPV to equal $0 (zero)? How should the firm use this answer to decide whether they should implement this solution?

d. If you had included the full $20 million as a fixed cost in the model, why would the model not have recommended closing the two warehouses and opening the new plant?

7. Open the file `Accounting Allocations.xls` located on the book Web site to see a list of sample accounts that a firm has for their various plants. Your team has already gone through these accounts and has determined whether the costs are fixed or variable. What are the fixed and variable costs you should load into the model based on this analysis? Which plant has the highest overall fixed cost? Which plant has the lowest per-unit variable cost? How would you rank the competitiveness of these plants?

8

BASELINES AND OPTIMAL BASELINES

Up to this point, we have compared the results of our optimization run to other optimization runs. However, the goal of most network design projects is to improve an existing supply chain. For example, when you are presenting your results, you may be asked how much money is saved with the new design or how much the service improves. To answer these questions, you need a baseline. A baseline is a model of the existing network.

Equally important, you will also be asked a deeper question: How do you know that your model is valid? The baseline model can solve that problem as well.

It turns out that there are two main baseline models you will want to consider. Each provides its own set of insights. We will call the two models the *actual baseline* and the *optimized baseline*. Both are important and both are critical.

Actual Baseline

The actual baseline is a representation of the supply chain exactly as it was run in the past. In practice, most modelers use data from the previous year to build the actual baseline model. In the actual baseline, you have exactly the same mathematical formulation that we've previously covered except that you are eliminating any choices for the model.

That is, the actual baseline uses just the existing facilities and the current assignments of facilities to customers (in more complex models you will extend this for flows between facilities and where product is made and how much is made at each location). In terms of the mathematical formulation, we are taking the decision variables and assigning them a value. During an optimization run, we have let the optimization engine pick the best value. Now we are telling the model which facilities are open and which ones are closed. We are telling the model which customers receive product from each facility. And so on. In effect, you are turning the mathematical model into a big calculator.

When you are building this baseline, it is important that you set up the model exactly the same way you are going to set up the optimization runs. That is, if you are going to run the optimization with outbound transportation costs, you want the baseline to include the outbound transportation costs.

We have found it helpful to set up the baseline in two steps. In the first step, you set up the model just as we set up the optimization models in previous sections. That is, you set up the structure of the model first. It is this structure that you will use later when running optimization scenarios.

The second step is to lock in all the decisions. In this step, you input data such that all the existing facilities are used, and you specify the flow from each facility to each customer. In previous chapters, our maps and solutions all looked relatively clean. In the actual baseline, we do not expect this to happen. Here you are entering data as it happened. For example, in the actual baseline, you may have shipments from your New York facility to your customer in Los Angeles even though you have a facility in Los Angeles. Things like that happen all the time in a real-world supply chain. We do not want to pretend they don't.

After you have this model built, it provides you a nice way to validate the structure of the model and the costs. The idea is that your baseline model reflects the business you are modeling. You compare the results from the model with the costs that the company actually incurred. You typically want the costs of your model to come to within 1% to 10% of the actual costs. There is always a good discussion on how close you need to get to the actual costs, and a lot of it depends on the culture of the company you work for and the importance of the project.

The main argument for getting close to 1% is the fact that the firm is making important decisions and they want to feel comfortable that the model that is guiding the decisions is accurate. Because you are working with historical data, it should be possible to match the model costs with historical costs. If you cannot, there may be a problem with the model. In this case, you do not want to fall into the trap of fooling yourself that costs within 1% guarantee that your model is fine. There are ways to load a baseline to get the costs you want that could make future optimization runs misleading.

The main arguments for being fine with something in the range of 10% can be two-fold. First, the availability of quality data may be an issue, thereby requiring the use of assumptions to fill gaps in data. For example, you may be missing some data on outbound shipments to customers but have information on total freight spend. In this case, you can input outbound data that is available and allow the model to determine the remaining customer assignments through optimization. These optimized assignments may be slightly different from what actually happened, thereby yielding a baseline cost that could be 10% away from actual costs. This is not necessarily a bad thing as we will be generating a similar result with the optimized baseline (discussed below) which will be used extensively during the analysis.

Secondly, the firm realizes that a supply chain is complex and always changing, so it is difficult to model costs exactly. And the firm is not going to make a decision unless the savings are relatively large compared to the baseline. So the baseline serves as an anchor, and if we find 20% savings from the baseline, we can assume that we've found a signicant savings. In this case, you still want to make sure that the baseline model is functionally accurate and not off by 10% because of mistakes.

Besides comparison of the overall cost of the model to the overall actual costs, you also want to compare the costs at a detailed level. For example, if the actual transportation costs are $10 million, you want to come close to the $10 million. And if the transportation costs from one warehouse are $3 million, you want to come close to that number as well. You can continue this process for other variables and at a finer level of detail.

This detailed analysis not only tests the cost validity, but also helps test the structure of the model. That is you, are testing the underlying details to make sure the model is behaving correctly. You can also make small tweaks to the baseline to make sure that the costs move in the direction they should. This also tests the structure, and we'll explore this more in the optimized baseline section.

The reader may ask: which project has better results, one that ensures a 1% difference or 10% difference between the model and reality? The answer is that we have seen both types of projects be successful. You will have to use your best judgment for each project. Also remember, an important part of these studies is selling the results internally. A big part of selling internally is convincing people that the model is correct. In some firms, you will need the baseline results to be very close to actual results, and in other firms not as close.

After you have validated the results, you now have a reference point and a model that has passed at least one check of its validity. But, keep in mind, that even though you have used the actual costs to validate your model, this is only one test. You need to keep a vigilant eye on the model validity throughout the entire project.

This baseline model needs to be ready to be converted to the optimization model for two reasons:

1. You need a fair point of comparison. If you change the structure of the model before running the optimization, you can no longer be sure your comparison is fair.

2. You need a valid starting point. The actual baseline has been validated. If you change the structure, you may no longer have a valid model.

It is good practice to minimize the changes to the structure of the model in going from one scenario to another. If you change the structure, you really need to revalidate the model.

Optimized Baseline

After the actual baseline is done, you will want to run the optimized baseline. In the actual baseline, we modeled everything that actually happened. In the optimized baseline, you want to replicate what should have happened based on the rules you had in place and if you executed according to plan. In our example from earlier, we want the Los Angeles customer to receive product from the Los Angeles facility, not the New York facility. You can also think of the optimized baseline as "cleaning up" the actual baseline.

An optimized baseline is not as well defined as an actual baseline. You have some flexibility in how you set up the optimized baseline model. And you may want to run several versions of the optimized baseline. The spirit of the optimized baseline is to clean up the baseline and come up with a model of what should have happened. Typically, an optimized baseline has the following structure:

- Uses all the existing facilities
- Uses existing assignments of facilities to customers (or later from one facility to another). In some versions of the optimized baseline model, you may relax this constraint and let customers receive product from different warehouses.

So, what does this model tell us?

First, it helps us validate our model. We want the results of this model to make sense. When we clean up the model and let products flow as we planned, are costs reasonable, are capacities still respected, and are other business rules respected? This reminds us that we need to continually evaluate our results. Real-world supply chains are complicated. Models are just a representation of the real world. We need to keep a skeptical eye on our models to make sure that they are giving results that reflect enough of the real world to allow us to make decisions.

Second, the model helps us identify areas for improvement. If there is a significant and validated cost difference between the actual baseline and the optimal baseline, it may signal a chance to improve. In this case, you want to understand why the actual supply chain deviated so far from the rules you put in place. You may have an opportunity to change the way different parts of the business operate so that you can better follow the rules that are in place. In other words, if you can execute better, there may be significant savings opportunities. Although no changes to the physical infrastructure of the supply chain are needed, you can still find significant savings. Occasionally, there is more benefit in running the existing supply chain better than re-designing it.

Third, this model provides a good benchmark to compare our optimization runs. When we run an optimization model, it will not make illogical decisions (it won't ship from New York to Los Angeles if there is a capable facility in Los Angeles). So we do not want to give ourselves a false sense of the results by comparing an optimization run to an actual baseline. That is, the actual baseline may have a high cost based on poor execution of the plan and not a poor plan. By comparing to the optimized baseline, we are comparing optimized scenarios versus an optimized version of the baseline so we have a fair comparison.

Other Versions of the Baseline

In some cases, the baseline and the optimized baseline as we described previously may not fit your situation completely. The following two cases provide immediate challenges:

1. The supply chain you are modeling experienced significant changes in the past year.

2. The supply chain you are modeling does not have a logical baseline or historical set of data.

Case #1 can happen if the firm has opened and closed significant facilities in the past year or has changed the product mix in a dramatic fashion. This means that the actual costs from the past year are not going to be valid and may not provide a clear basis for validating our baseline model.

One way to accommodate this is to run two or three different baseline models. The baseline models would capture the cost of what would have happened if the changes hadn't occurred, the costs of what would have happened if the changes had happened at the start of the year, and the combined cost that attempts to match up the costs with what actually happened. Then, going forward in the analysis, you want to use the model with the changes as your starting point.

As an example, let us say that a firm shut down a plant in July last year, and we are looking to use data for the last calendar year. With a planned plant shutdown, you would expect to see production going down at this plant in the months leading up to July, and consequently volume picking up at other plants through July and beyond. In this situation, we can create two baselines—one representing January to July, with the old plant showing production, and another baseline for August to December, representing the network without the shut-down plant. After we validate both models using their respective data, we can then use the second baseline model (August to December) as the basis going forward as this is the new reality. For purposes of consistency, we can take this baseline and annualize its volume and use this version (call it revised baseline) to apply demand growth forecasts for scenario analysis.

Case #2 can happen if you are modeling an entirely new supply chain or are modeling so far out in the future that last year's demand is not considered a valid starting point. This case is a little trickier and requires some creativity.

The first thing to keep in mind is that you need to build some baseline. You need a way to validate the structure of the model and validate that the costs are reasonable. In some cases, the firm will have a preliminary budget in mind and you can use that as your reference point. Or you can use a version of the optimized baseline as the baseline. That is, you determine the cost of the new business using your existing structure. There is no correct answer here, but the more similar you can make the baseline to your existing network, the easier it will be to validate your model.

Baseline Case Study—Illinois Quality Parts, Inc.

Let's go through a case study to learn and understand how to develop baseline and optimized baseline models before starting to evaluate alternatives. We will take the case of Illinois Quality Parts, Inc. (IQP), a large distributor of industrial parts and components based in the Chicago area. The company distributes a wide variety of industrial parts ranging from small components such as bearings and belts to heavy parts used by original equipment manufacturers (OEM) that make of automotive, heavy equipment, and industrial products.

The company distributes these products through their network of three warehouses located in Riverside, California; Addison, Illinois; and Bridgewater, New Jersey. The parts are procured from specialized manufacturers of the parts located all over the U.S. The parts are delivered to the three warehouses with freight paid by the vendors.

The customer base includes original equipment manufacturers, small distributors, and retailers located across the U.S. The company consolidates customer orders and ships them in full truckloads to its customers. Each warehouse has a customer territory assigned to it, based on which, the customer orders are routed to the appropriate warehouse and shipped out. The map in Figure 8.1 shows the customer territory by warehouse (or called a DC-distribution center by IQP).

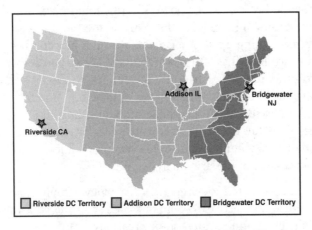

Figure 8.1   Customer Territory Assignment by Warehouse

Looking deeper into their supply chain costs, the team calculated that they were spending $78 million in outbound transportation costs annually. See Figure 8.2 for a breakdown of outbound freight costs by warehouse. (Note that the totals may not add up due to rounding.)

Warehouse	Total Weight Shipped (Million Lbs)	Outbound Freight Cost ($MM)
Addison, IL	1,091	$34.85
Bridgewater, NJ	1,205	$26.34
Riverside, CA	604	$16.35
TOTAL	2,900	$77.55

Figure 8.2   Annual Outbound Freight Costs by Warehouse

The management team had not performed a network analysis in over five years and wanted to understand whether there were opportunities to reduce costs—this included evaluation of customer assignments to warehouses, and changing the number and location of warehouses in the network.

Analyzing Shipment Data and Creating Customers and Demand

Before we set out to answer questions around alternatives, first we want to build a baseline model that accurately represents how the supply chain operates today. To build such a model, we will want to collect and use actual historical data as this will show how the company's supply chain actually worked.

We will start our analysis by collecting historical shipment data and building the customers and demand. Because this analysis will primarily focus on transportation costs, we will use weight (pounds) as the unit of measure to model customer demand. From a transportation perspective, the vendors pay for inbound freight, and this will likely not change with any network changes; so the management team wanted to focus on only the outbound costs.

A sample portion of the shipment data is shown in Figure 8.3. This data serves two purposes: (1) We can use this to derive the demand by customer for the model, and (2) this tells us which warehouse shipped to which demand point over the past year.

Shipment Number	Origin Loc	Origin City	Origin State	Destination City	Destination State	Dest Zipcode	Shipment Weight (Lbs)	Freight Cost
1035	Addison DC	Addison	IL	Nashville	TN	37211	39,535	$920
1036	Addison DC	Addison	IL	Cedar Rapids	IA	52402	40,053	$457
1037	Addison DC	Addison	IL	Arlington	TX	76063	38,545	$1,890
4542	Riverside DC	Riverside	CA	Salem	OR	97301	39,421	$1,932
4543	Riverside DC	Riverside	CA	San Diego	CA	92154	37,033	$450
5125	Bridgewater DC	Bridgewater	NJ	Johnson City	TN	37660	40,130	$790
5126	Bridgewater DC	Bridgewater	NJ	Albany	GA	31705	38,053	$1,296
5127	Bridgewater DC	Bridgewater	NJ	Worcester	MA	01604	39,460	$690

Figure 8.3   Sample Extract from Historical Outbound Shipment Data

We then take the shipment data to identify the customer locations (from the Destination fields in the shipment data) and the total demand (sum of shipment weight) for the baseline model. To create the list of the customers, we group all the same destinations together. That is, the shipment file may have 100 line items for shipments to the same location and these will be grouped into a single customer. Then, we add up the total demand for each of these customers. Figure 8.4 shows the customers plotted on a map and the states shaded by their relative demand.

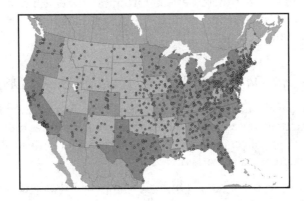

Figure 8.4   Ship-to Customer Locations and Demand Distribution by State

Modeling Historical Data as Predefined Flows

We can see from the sample data (in Figure 8.3) that customers in Tennessee (TN) were served from both Addison and Bridgewater warehouses, even though Tennessee falls under the Addison warehouse territory according to their customer assignment rules.

Why would this happen? This is very typical of any given supply chain as there are several reasons that require departure from the published or expected policy. The most likely reason for out-of-territory shipments is inventory availability, wherein there may not be sufficient inventory of the ordered items to ship from the assigned warehouse, thereby requiring shipment from another warehouse. In these cases, it is cheaper and faster to ship to the customer from another warehouse than to get the product replenished at the Addison warehouse from Bridgewater warehouse and then ship to the customer.

We will revisit this topic after we run the baseline model to understand the magnitude of these out-of-territory shipments.

So, our next step is to replicate these shipments in our model. For example, we want to show customer demand in Tennessee being served from both Addison and Bridgewater warehouses. We can model this by predefining the flow of product in each lane based on the shipment data. Each commercial network modeling software has its own way of inputting this data into the model, but the general concept is to specify the total volume (in pounds in this example) moving from each warehouse to each customer.

We see an example of the predefined flow for the Nashville, Tennessee, customer in Figure 8.5. Based on the shipment data, the total demand for this customer is 4,259,050 pounds; we are predefining that 2.981 million pounds of this demand would be served from the Addison warehouse, and the remaining demand (1.27 million pounds) would be served from the Bridgewater warehouse. We then fix the variable to honor these constraints. This concept of predefined flows is applied across all the ship-to locations in the baseline model.

Source Warehouse ID	Source Warehouse	Destination Customer ID	Destination Customer ▲	Min Flow (lbs)
1	Riverside, CA	100337	Nashville - TN	0.00
2	Addison, IL	100337	Nashville - TN	2,981,340.00
3	Bridgewater, NJ	100337	Nashville - TN	1,277,720.00

Figure 8.5   Example of Predefined Flow for Baseline Model

Modeling Transportation in the Baseline

Now that we know the flows, let's add the transportation costs. The shipment data file included shipment weights and actual freight costs by truckload shipment. We can use this to derive transportation cost per mile rates for the model using regression analysis (as described in Chapter 6, "Adding Outbound Transportation to the Model").

We do not yet know the right level of granularity that would provide an acceptable level of accuracy to the baseline model—for example, will it be acceptable to calculate and apply a $/mile rate across all lanes, or do we need to estimate the rates at some lower level?

As mentioned before, we typically aim for getting the baseline model costs to within 1% to 10% of actual costs. The higher end of the range is typically acceptable if the quality of data was poor, requiring the use of assumptions to substitute real data. If our modeling analysis yields a baseline model that falls within this tolerance and gives a reasonable level of confidence that it is accurate, we have met our goals for this task.

To go through this process, we will start modeling transportation with an overall $/mile rate across all lanes. If this does not yield a sufficiently accurate result, we will recalculate at a lower level and reapply in the model.

The Logistics team at IQP estimated that their average truckload rate was $2/mile across all lanes, with a minimum charge of $450/load. Based on a quick regression analysis using all the lanes as the sample, we come up with something close to this as well. We will start with using this $2/mile rate in the baseline model.

Baseline Model Results

The maps showing results of the baseline model are displayed in Figure 8.6. The maps from the baseline model depict the reality of how customer demand was met. There were out-of-territory shipments from all three warehouses, including products moving from the Bridgewater warehouse all the way to California.

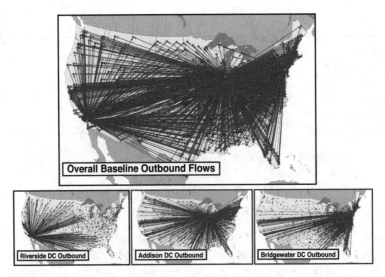

Figure 8.6   Maps Showing Aggregate Baseline Model Results and by DC

Now let's look at what the baseline transportation costs look like relative to the actual freight spend, based on the $2/mile TL rate (see Figure 8.7).

Warehouse	Total Weight Shipped (Million Lbs)	Outbound Freight Cost ($MM)	Baseline with $2/Mile TL for All Lanes	
			Model Cost	Delta vs Actual
Addison, IL	1,091	$34.85	$35.35	1%
Bridgewater, NJ	1,205	$26.34	$32.94	25%
Riverside, CA	604	$16.35	$15.22	-7%
TOTAL	2,900	$77.55	$83.51	8%

Figure 8.7   Results Comparison—Actual Versus Baseline Transportation Costs

When we compare the total transportation costs between actual ($77.55 million) versus the baseline model ($83.51), it shows that the model results were about 8% different from the actual costs. This is actually not a very bad result and can be acceptable in a lot of cases. Before we make a decision on whether this is acceptable, let us dig a little deeper at the cost comparison by warehouse. The costs out of Addison are very close to actual (1%), but the freight costs for Bridgewater outbound are off by 25% from actual costs, while Riverside outbound costs are off by 7%. The 7% delta can be ignored for now, but the 25% delta is very large. What does this mean? The comparison shows that the $2/mile is roughly 25% higher than actual for New Jersey origin. This is important to note—all things being equal, the model will assign customers to a warehouse that provides the lowest cost per unit, which is a factor of both distance and $/mile. If the rate is lower than $2/mile, that means the New Jersey warehouse can potentially serve customers farther away than previously estimated. It could also affect the choice of warehouse location picked because it may be cheaper to locate a warehouse in a zone or state that has lower outbound rates.

Based on this previous analysis, we arrive at the conclusion that a generic $2/mile rate will not be appropriate for all lanes. The logistics department was able to provide a state-to-state matrix with $/mile rates that we can use in the model. See Figure 8.8 for a sample extract from the matrix.

FROM \ TO	AL	AR	AZ	CA-N	CA-S	CO	CT	DE	FL-N	FL-S	GA
AL	$2.01	$1.73	$1.66	$1.49	$1.45	$1.89	$2.00	$1.93	$2.48	$2.48	$1.70
AR	$1.70	$2.06	$1.65	$1.53	$1.51	$2.11	$1.86	$1.72	$2.04	$2.13	$1.57
AZ	$1.23	$1.29	$1.76	$1.28	$1.22	$1.68	$1.51	$1.33	$1.46	$1.53	$1.26
CA-N	$1.23	$1.23	$1.80	$1.74	$1.60	$1.86	$1.55	$1.32	$1.45	$1.61	$1.24
CA-S	$1.30	$1.27	$2.09	$1.93	$1.72	$1.94	$1.55	$1.40	$1.48	$1.62	$1.30
CO	$1.16	$1.16	$1.39	$1.25	$1.17	$1.88	$1.36	$1.31	$1.43	$1.65	$1.19
CT	$1.06	$1.08	$1.31	$1.34	$1.31	$1.68	$2.18	$1.63	$1.49	$1.71	$1.09
DE	$1.12	$1.12	$1.34	$1.34	$1.27	$1.63	$2.26	$2.42	$1.56	$1.78	$1.16
FL-N	$1.09	$1.14	$1.40	$1.34	$1.33	$1.60	$1.65	$1.54	$1.94	$2.38	$1.19
FL-S	$0.88	$0.94	$1.25	$1.28	$1.25	$1.31	$1.65	$1.36	$1.80	$2.09	$0.92
GA	$1.72	$1.52	$1.52	$1.43	$1.38	$1.74	$1.99	$1.91	$2.46	$2.48	$2.06
IA	$1.57	$1.68	$1.63	$1.51	$1.46	$1.97	$2.02	$1.89	$1.84	$2.06	$1.63
ID	$1.29	$1.29	$1.53	$1.36	$1.23	$1.74	$1.50	$1.31	$1.43	$1.55	$1.29

Figure 8.8   Extract of Truckload Matrix

These types of rates typically have large states such as California and New York broken into North–South or East–West, given the differences in demographic and economic activity within the state.

Let's rerun the baseline model substituting the $2/mile rate with these detailed rates. Note that we are changing only the transportation rates; the predefined flows are not changing, so the maps for this revised baseline will look exactly the same as the previous iteration.

The results (as shown in Figure 8.9) show that the baseline model costs are much closer to actual with the state-to-state level rates. The total cost is within 3% of actual and within 5% of actual by warehouse. This tells us that we are reflecting the differences in outbound rates by origin and potentially by destination, thereby yielding model costs that are very close to actual.

Warehouse	Total Weight Shipped (Million Lbs)	Outbound Freight Cost ($MM)	Baseline with $2/Mile TL for All Lanes		Baseline with State-to-State Rates	
			Model Cost	Delta vs Actual	Model Cost	Delta vs Actual
Addison, IL	1,091	$34.85	$35.35	1%	$36.60	5%
Bridgewater, NJ	1,205	$26.34	$32.94	25%	$27.71	5%
Riverside, CA	604	$16.35	$15.22	-7%	$15.83	-3%
TOTAL	2,900	$77.55	$83.51	8%	$80.14	3%

Figure 8.9   Results Comparison—Actual Versus Baseline with New Rates

Based on these results, the management team can feel comfortable that our baseline model is calibrated appropriately and is a good representation of how their supply chain operated historically.

Building the Optimized Baseline(s)

Now that we have a validated baseline model, our next step is to create and run an optimized baseline scenario. This scenario is intended to show what it would look like if the current network operated optimally and without any operational exceptions. In this case, we can come up with two such versions of optimized baseline:

1. Optimized baseline with customers served according to their assigned territories

2. Optimized baseline with optimized assignment of customers to warehouses

The first version will show what the cost would be if there were no out-of-territory shipments based on the existing territories. Because most of these exceptions are attributed to inventory availability, it will help quantify the financial impact of not having the right inventory in the right place at the right time.

In other words, the combination of baseline and optimized baseline models can help quantify the financial impact of these operational issues, and help the team initiate immediate corrective action.

The second version of the optimized baseline will tell us whether the customer territory assignment itself can be improved—that is, whether there are opportunities to reassign specific customer states or locations among the warehouses to reduce overall costs. The customer territories have not been evaluated by the logistics team in a few years, so it may identify quick opportunities.

Note that in both scenarios we are not making any changes to the warehouses or the network itself. We will keep the warehouses fixed and not look at any new locations in these scenarios—these will come later after the optimized baselines are completed.

The results of the two optimized baseline scenarios are shown in Figures 8.10 and 8.11. The map for the scenario with current territory assignments shows flows as expected according to the current business rules. The map for the scenario with optimized territories shows the Riverside warehouse picking up states such as Colorado, New Mexico, and Montana. We also see Bridgewater shipping to Southern Louisiana and Southern Mississippi, which were originally part of the Addison territory.

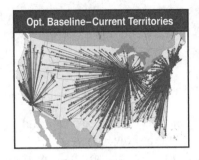 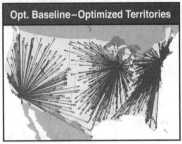

Figure 8.10   Maps Showing Results of Optimized Baseline Scenarios

Metric	Baseline	Optimized Baseline - with Current Territories	Optimized Baseline - Optimized Territories
Transportation Costs (in millions)	$80.14	$77.51	$74.39
Savings over Baseline	-	$2.63	$5.75
% Savings	-	3%	7%
Avg. Distance to Customers (miles)	535	504	504
Reduction over Baseline		31	31
Volume by Warehouse (in mm lbs)			
Riverside	604	583	731
Addison	1,091	1,173	845
Bridgewater	1,205	1,142	1,323
TOTAL	2,900	2,900	2,900

Figure 8.11   Results Comparison—Baseline Versus Optimized Baseline Scenarios

The cost and volume comparison provides us with a lot of insight on improvement opportunities.

This analysis shows that there is a *2.6 million or 3% cost reduction opportunity* if the outbound orders were shipped according to their assigned territories. This means that the company could save over $2 million if they had the right inventory at each ware-house—this is usually attributed to an insufficient amount of safety stock buffer at the warehouses. The company can use these cost savings versus the cost of higher inventory levels at these warehouses to determine what makes sense.

This scenario also shows that Riverside and Bridgewater are currently handling more volume (as represented in the Baseline) than they should based on their customer terri-tory assignments (Optimized Baseline—With Current Territories). The converse applies to Addison, which is serving less volume than it should be according to the territory assignments.

When we look at the second optimized baseline (with optimized territories), it shows that the company could *save $5.75 million or 7% over the baseline model* if the territories were realigned optimally. In this case, Riverside should be serving states farther east of its original territory, as well as other minor reassignments between Addison and Bridgewater warehouses as noted earlier.

The consequence of this realignment is that the Bridgewater and Riverside warehouses will handle additional volume. Note that we have not modeled capacity in this model so far, so this is not factored into these outputs. We can easily add the capacity constraints and rerun the model. However, there is value in running this scenario without capacity constraints—it tells us what the optimal throughput would be for each warehouse. This is valuable information that the company can use to determine whether this can be executed operationally.

As discussed in Chapter 5, "Adding Capacity to the Model," it is usually hard to precisely define warehouse capacity, so it is typical to run some scenarios ignoring capacity and evaluating the model outputs versus baseline (or current) in order to understand feasibility. In this case, the model recommends that Riverside handle 21% more volume and that Bridgewater handle 10% more compared to last year. The management team needs to evaluate whether this is feasible given current capacities. Based on past experience, we have seen that a 10% increase in volume can be easily accommodated even within warehouses that are at full capacity; a 20% increase is usually a bit more challenging if capacity is tight, so this needs further analysis.

Though the primary focus of this case study was the baseline and optimized baseline models, we will wrap up this case by showing the best two-, three-, four-, and five-warehouse solutions. Figure 8.12 and Figure 8.13 show how these solutions compare to the baseline.

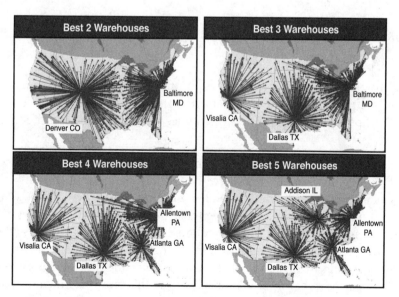

Figure 8.12   Results of Scenarios Picking Best Two, Three, Four, and Five Warehouses

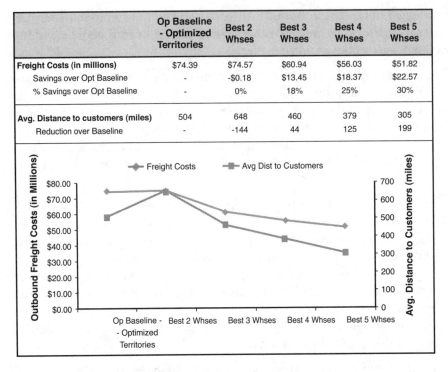

	Op Baseline - Optimized Territories	Best 2 Whses	Best 3 Whses	Best 4 Whses	Best 5 Whses
Freight Costs (in millions)	$74.39	$74.57	$60.94	$56.03	$51.82
Savings over Opt Baseline	-	-$0.18	$13.45	$18.37	$22.57
% Savings over Opt Baseline	-	0%	18%	25%	30%
Avg. Distance to customers (miles)	504	648	460	379	305
Reduction over Baseline	-	-144	44	125	199

Figure 8.13   Results Comparison of Scenarios with
Best Two, Three, Four, and Five Warehouses

Lessons Learned from Baseline and Optimized Baseline Modeling

The actual baseline model is a representation of the current supply chain and how it operated in the past. This is an important first step in the network modeling process because this helps validate that the network model accurately represents the supply chain and its flows.

After the baseline model is validated, it is important to run the optimized baseline scenario(s). This represents the current network and its locations but shows the impact if everything happened according to the current business rules. There can be several variations of the optimized baseline, including those in which some of the rules are relaxed (e.g., customer assignments to specific warehouses). This model also serves several purposes. It shows the potential improvement opportunities within the current supply chain without having to make major infrastructure changes. It also provides a good basis for comparing what-if scenarios—because this scenario includes optimized outputs, it serves as a good apples-to-apples comparison with optimized outputs from what-if scenarios. The optimized baseline can also help validate the model itself well.

A well-designed and well-developed baseline and optimized baseline model serves as a strong foundation for development and running of what-if scenarios.

End-of-Chapter Questions

1. What is the difference between an actual baseline and an optimized baseline?

2. Open the file from Chapter 3, "Locating Facilities Using a Distance-Based Approach," called MIP for 9-City Example.xls on the book Web site. Assume that the baseline had St. Louis serving itself, Chicago, and Cincinnati while Boston served all the other locations. How would you set this up in this model and what is the baseline solution? If, in the baseline, Atlanta, New York, and Charlotte were served by both St. Louis and Boston with a 50% split, how would you set up the baseline now and what is the solution?

3. If your baseline model's costs are 20% lower than the actual costs, under what conditions could this still be a valid baseline?

4. If your baseline model's costs are within 0.1% of the actual costs, under what conditions might this still not be a valid solution?

5. In the Illinois Quality Parts case study, we first ran the baseline model with a rate of $2/mile for all lanes. Figure 8.14 shows the output of this scenario showing total cost and total miles traveled out of each warehouse.

Warehouse	Total Weight Shipped	Baseline with $2/Mile TL	Total Miles (in Millions)	Avg $/Mile
Addison, IL	1,091	$35.35	16.97	$ 2.08
Bridgewater, NJ	1,205	$32.94	15.06	$ 2.19
Riverside, CA	604	$15.22	6.73	$ 2.26
TOTAL	2,900	$83.51	38.76	$ 2.15

Figure 8.14   Baseline Cost Comparison

a. Even though we applied a rate of $2/mile, the table (total cost / total miles) shows a rate higher than $2/mile for all warehouses. Why is that? (Hint: Think about the fact that the true transportation rates include minimum charges.)

b. The solution with the best two warehouses shows an average distance to the customers that is 144 miles higher than the optimized baseline with optimal customer assignments. However, the freight costs for the two scenarios are very close (~$74 million). How can you explain this?

c. When we look at the results of the scenarios shown in Figure 8.10, it looks as though some customers are not assigned to their closest warehouse. Why is that? Does that make sense?

6. You are starting a network design study for a company that wants to reduce costs in its distribution network. Their financial year is the same as the calendar year. Last May, the company changed their packaging strategy and the gross unit weight of their products increased by 20%—that is, for the same number of units ordered, the total weight shipped went up by 20% due to increased weight of packaging material. The products were shipped in full truckloads that typically weighed out. The demand is fairly stable all year with no seasonality. What timeframe would you use to pull data for the baseline? If you used all 12 months, what would be the advantage? What would be the disadvantage? How will you model demand for the scenarios?

9

THREE-ECHELON SUPPLY CHAIN MODELING

Up to this point, we've considered only a two-echelon supply chain—a facility and the customers it services. You have seen how optimizing the two-echelon supply chain can address many types of problems. However, by adding an echelon, we can capture yet another important trade-off in supply chain modeling: Should your facilities be closer to the source of the product or to the destination?

One type of three-echelon supply chain may include a set of plants or suppliers that ship to warehouses, and then the warehouses, in turn, ship to customers. Alternatively, we may consider a supply chain in which a group of raw material suppliers ships to a plant, and the plant then ships to the customers. In fact, after we learn about the key trade-offs in a three-echelon supply chain, we may quickly find it useful and easy to add a fourth, fifth, or even more echelons to the model as well. The model becomes more complex with each additional network tier considered, but the same fundamental trade-offs still exist. Let's now look more closely at a case study that will help us understand and learn about modeling and analyzing three-echelon supply chains.

JADE's Corporate Background

Walter Jade started the JADE Paint and Covering business in Suffolk, New York, in 1906. Over a hundred years later, the company has expanded across the U.S., but their main distribution center still remains in the city where it all started. (Figure 9.1 depicts this warehouse as a triangle in eastern New York.) JADE originally started manufacturing only paint, but has since expanded their product line to include specialty paints, wood treatments, ready-mixed concrete for patching, and several other related products.

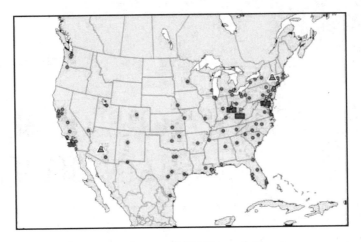

Figure 9.1　Current JADE Network Structure

In the markets they serve, they have strong brand recognition and professional contractors appreciate of the quality of the product. This allows JADE to command premium prices and to expect loyalty from their customer base.

Walter Jade started the business selling to customers in the Northeast but quickly expanded to most of the eastern seaboard and several Midwest markets as well. As previously mentioned, however, the company's main distribution center to service these markets still remains in upstate New York.

These original markets had been good to JADE over its long history, but within the past ten years JADE decided to make a few key acquisitions. This has allowed them to break into many additional markets on the West Coast and in the greater Texas area. You can see all their current markets displayed as dots on the map in Figure 9.1.

Although these new markets have helped grow revenue, they have proven to be quite costly from a transportation point of view. JADE quickly made the decision to open a new warehouse location in Phoenix, Arizona, to help handle these markets and alleviate some of their costly outbound transportation.

JADE has three plants in the U.S. and one in China. All products from China come into the U.S. through the port of Long Beach in California. The three plants and the port of entry for China products are displayed as rectangles with small flags on top on the map shown in Figure 9.1. All JADE manufacturing plants are known in the industry for having low costs while still producing extremely high quality product.

JADE's products can be categorized into four basic product families. Currently, each plant specializes in making just one of the product families. Therefore, each JADE product family is produced from just one of their four plants.

The JADE plant in China was the result of an acquisition and is not considered to be an optimal source of the products it specializes in producing. JADE realizes how unusual it is for such a heavy product to be made so far away from its market. Competitors have always raised an eyebrow at this decision, and JADE management continually questions whether they should move production closer to the market.

Transportation-wise, the plants and warehouses are set up to handle rail shipments. Therefore, shipments from the plants to the warehouses are a mix of both truck and rail transit. Outbound to customers, product travels in mostly full (but not always) truck-load shipments. Last year, JADE did some quick transportation analysis and determined that their shipments from a plant to a warehouse cost approximately $0.07 per ton-mile, and their cost of shipments to their customers was close to $0.12 per ton-mile. In both cases, the rate structure has a minimum charge of $10 per ton. Because of this minimum charge, if a 20-ton outbound shipment from a warehouse to a customer travels only a short distance, the cost will default to $200 for that load.

For more details on this case study and a look at the JADE network see the supplemental material titled `Jade Intro Case.zip`, located on the book Web site.

Now that we have been introduced to JADE and have been given a good review of their network, let's learn more about what has triggered the company to want to optimize their current supply chain.

Over the past year, JADE has seen their profits start to fall and has begun to look closely at their current spend in all areas of the company. The CEO is worried that the whole firm may go out of business if it cannot reduce supply chain costs. This newfound uncertainty led, in part, to the resignation of their former VP of the supply chain, and his replacement is now under intense pressure to take costs out of the supply chain immediately. If he cannot find ways to remove costs, he will likely be removed from his new job in no time at all.

The new VP of supply chain is confident, however, that JADE can remove a significant amount of transportation costs from the network with a more optimal network configuration. Remember that in the current supply chain, JADE is serving its entire market from just the two warehouses previously mentioned. A map of this baseline structure is shown in Figure 9.2.

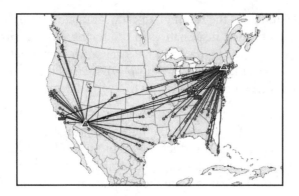

Figure 9.2   JADE's Current Network Solution

JADE's current transportation spend is $254 million per year, not including the additional cost to ship product from the plant in China to the port of Long Beach. Of this, $133 million is the cost to get product from the plants to the warehouses and $121 million is the cost to get product to the customers.

The new VP is convinced that they can reduce these numbers in the future. A quick look at the current state map shows that both of their warehouses serve very large territories for such a heavy class of products. In fact, a detailed analysis shows that only 21% of the demand lies within 200 miles of their current warehouse locations. Furthermore, only 32% of demand is within 400 miles.

Before we solve the JADE case study, let's explore this type of problem in general.

Determining Warehouse Locations with Fixed Plants and Customers

To help us analyze JADE's case, we will start with the simplest and most common three-echelon problem: locating warehouses given fixed locations of plants and customers. A model with a set of fixed customers is nothing new; all models in our previous examples had this basic assumption. Now we want to add a set of plants to the model where product is produced. Each plant may make different products or the same set of products. To start, we want to determine the best number and location of warehouses to minimize cost.

In this model, the plants ship to the warehouses and then the warehouses ship to the customers. We will assume that the plants cannot ship directly to the customers. This is very common in practice. Shipping direct to the customer requires a different set of systems and processes than shipping from a plant to a warehouse. For example, most plants are set up to produce large batches of a product. After products come off the production line, they are packed on to large pallets which are loaded into full trucks and shipped to a warehouse. When shipping to customers, a facility would need space to assemble the orders including the ability to put different products made by different plants on to a pallet or case for the customer. The facility would also need the ability to ship to its customers in various modes such as truckload, less-than-truckload, and parcel. Many plants do not have the space or capabilities for any of these requirements.

In addition, in many cases, the supply chain manager cannot consider changing the locations or capabilities of the plants. The plant locations are fixed and are not going to change. These may represent supplier locations that we cannot change, plants that required significant investment and are very difficult to move, or plants that require specialized skills and a trained labor force, or, quite simply, the company may not want to consider the relocation of the plants at this time.

The number and location of the warehouses is often an easier decision for many firms to make. And, plenty of savings opportunities are often discovered by relocating warehouses.

What makes this network design problem different from those in our previous chapters is that now the location of the existing plants or supply points will have an impact on the optimal location of the facilities as well. In the previous models, the facility locations were "pulled" to be close to the customers. That is, the model tended to locate facilities close to customers to reduce transportation costs. Now, we have two forces pulling on the location of the facility—the customer demand and the supply points. Simply stated, we want to minimize the cost of shipments both to and from the facility.

The relative strength of these two forces depends greatly on the problem being solved. In general, it comes down to the relative cost difference between the inbound and outbound costs. To highlight this, let's review our previous discussion on transportation costs.

Let's start by assuming that we have truckload (TL) shipments from our plant to the warehouse. That is, we easily fill trucks directly from the production line and ship these large quantities straight to a warehouse. This is a fairly common assumption. The warehouse then ships less-than-truckload (LTL) to customer points. That is, the customers order in smaller quantities that require this more-expensive transportation mode. Remember that on our cost-per-ton-mile estimates (the cost to drive one ton of product one mile), LTL transport is approximately three times as expensive as TL.

Let's understand how this impacts the optimal location of a warehouse. Take a look at Figure 9.3. In Case #1, the plant ships via TL a short distance to the warehouse. The warehouse then ships the product to the customers via a fairly lengthy LTL shipment. In Case #2, we have the same set of customers, but the warehouse is much closer to the customers. Therefore, the plant has a relatively long shipment to the warehouse but, in turn, a short shipment to the customers.

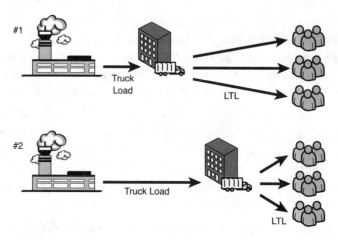

Figure 9.3   Optimal Warehouse Location Visual

Which supply chain would you guess has the lower cost? We are assuming that everything else stays the same between the two cases. That is, we have exactly the same number and size of shipments. We have simply changed the *location* of the warehouse.

In this example, Case #2 will have the lower cost. In this instance, we are taking full advantage of the fact that TL is only one-third as expensive per ton-mile and are shipping the product to a warehouse as close to our customers as possible. So we are minimizing the total cost by maximizing the distance we ship on TL and minimizing the distance we ship via LTL.

If we extend this concept further and try to minimize the distance on LTL by adding a second warehouse (as shown in Case #3 in Figure 9.4), have we lowered the cost even further? To keep it simple, we will stick with our assumption that everything is the same between the two cases except the number and location of the warehouses. The customers still get the same sized shipments and the plant has enough volume that it can ship in full truckloads in both Case #2 and Case #3.

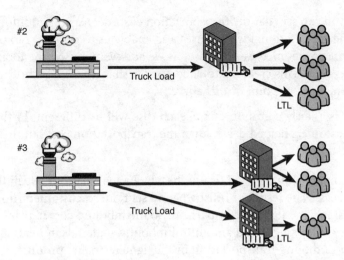

Figure 9.4   Optimal Warehouse Location Visual with Second Warehouse Option

Now, the answer is not as clear. For sure, Case #3 has lower transportation costs. But now we have a second warehouse. Presumably, this warehouse is not free. So now our answer to the question of which supply chain has the lower cost is "it depends." It depends on the cost of adding that second warehouse. If the transportation savings offsets the warehouse cost, Case #3 has the lower cost; if not, Case #2 is better.

If we generalize this case, we can represent this graphically as a trade-off curve shown in Figure 9.5.

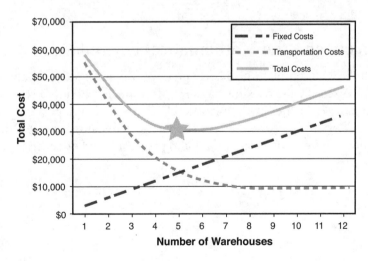

Figure 9.5   Number of Warehouses Versus Total Cost Trade-Off Curve

In this curve, you can see that the transportation costs decrease as additional warehouses are added. These additional warehouses also enable us to get closer to our customers and reduce transporation costs. However, as we add warehouses, the total fixed cost of the warehouses rises. This creates a total cost curve shaped like a bathtub. The best cost solution is at the bottom point of this curve.

Although this is a fairly common curve, each case will be different. In this case, being closer to the customers helped drive down the transportation cost but this is not always the case. For example,

- A plant may require a good deal of heavy raw materials for use in the production process. The resultant finished good sent out to customers, however, is relatively small and light in comparison. So the inbound cost may be expensive because of the weight and amount of inbound materials in relation to the cheaper outbound transportation of the lighter, smaller product.

- A retailer may ship full truckloads of a large assortment of its products to stores but source each product from a separate supplier requiring a lot of small shipments into its warehouse. As a side note, if these suppliers are widely dispersed geographically, they won't have much influence on the location of the warehouses since the pull of the suppliers in the east will cancel out those on the west, for example. And, retailers also tend to put a lot more emphasis on the service levels they are able to offer to their stores. So, while being closer to suppliers might save them in transportation, they often choose to ignore this and prioritize the service level they offer to the stores instead.

The Problem and the Mathematical Formulation

For the multi-echelon problem in this chapter, we will add onto the previously developed models (especially the one in Chapter 7, "Introducing Facility Fixed and Variable Costs"), including capacities at the plant level. There are many possible extensions—here is one such extension. Note that, for simplicity, we renamed some of the constants because we now have explicit "warehouses and plants" instead of generic "facilities."

$$\sum_{l \in L} \sum_{i \in I} (transPW_{l,i} + pVar_l)Z_{l,i} +$$

$$\sum_{i \in I} \sum_{j \in J} (transWC_{i,j} + whVar_i)d_j Y_{i,j} +$$

$$\sum_{i \in I} \sum_{w \in W} whFix_{i,w} X_{i,w}$$

Subject to:

$$(1) \sum_{i \in I} Y_{i,j} = 1; \forall j \in J$$

$$(2) \sum_{i \in I} \sum_{w \in W} X_{i,w} \geq P_{\min}$$

$$(3) \sum_{i \in I} \sum_{w \in W} X_{i,w} \leq P_{\max}$$

$$(4) \sum_{w \in W} X_{i,w} \leq 1 \forall i \in I$$

$$(5) \sum_{j \in J} vol_{i,j} Y_{i,j} \leq \sum_{w \in W} whCap_{i,w} X_{i,w}; \forall i \in I$$

$$(6) \sum_{l \in L} Z_{l,i} = \sum_{j \in J} d_j Y_{i,j} \forall i \in I$$

$$(7) \sum_{i \in I} Z_{l,i} \leq pCap_l; \forall l \in L$$

$$(8) Y_{i,j} \leq \sum_{w \in W} X_{i,w}; \forall i \in I, \forall j \in J$$

$$(9) Y_{i,j} \in \{0,1\}; \forall i \in I, \forall j \in J$$

$$(10) X_{i,w} \in \{0,1\}; \forall i \in I, \forall w \in W$$

$$(11) Z_{l,i} \geq 0; \forall l \in L, \forall i \in I$$

A few notes on the model:

- Because we now have warehouses and plants, we added facility-specific prepends to certain terminology: *whVar* (the warehouse variable costs), *pVar* (the plant variable costs), *transWC* (the transportation cost from the warehouse to customer), *transPW* (the transportation cost from the plant to warehouse), *whCap* (the warehouse capacity), *pCap* (the plant capacity) and *whFix* (the warehouse fixed cost).

- The objective function here divides into three distinct sections. The first section computes the total cost of producing or acquiring our goods, and shipping them to the warehouse. The second section computes the total cost of handling our goods at the warehouses, and shipping them to the customers. The third section computes the total fixed cost of opening the warehouses. One isn't obliged to organize the objective function in just this way. For example, we could have just as easily broken it out into five distinct sections, one each for production, plant-to-warehouse shipping, warehouse handling, warehouse-to-customer shipping, and warehouse fixed costs. So long as the correct total is computed, the objective function can be organized in whatever manner seems more readable and intuitive.

- The Z variable is new. This is the flow from the plants to the warehouses. This flow isn't constrained to be binary—it can take on any nonnegative value. This is the first continuous variable we have used thus far. Note that if a particular type of MIP solver is used, and if demand and capacity values are all integral, then this value will naturally take on integral values.

- Equation (6) represents a family of constraints that is new to this book, but storied in the annals of network flow modeling. These constraints are called "conservation of flow" constraints, and they will appear in any reasonably sophisticated shipping model. "Conservation of flow" constraints transmit the obligation to ship and produce from the "sink" points (the customers) to the "source" points (the plants or suppliers). Without them, the goods would simply materialize at the warehouses, as they do in our previous one-tier models. Because the whole purpose of developing the more complex two-tier model is to allow the solver to include the plant production and shipping costs in the optimization decisions, these constraints are crucial. These particular constraints state that for each warehouse, the total amount of goods inbound from the plants must equal the total amount of goods outbound to the customers. The left hand side sums the inbound shipments over all the plant-to-warehouse lanes that terminate at this warehouse, and the right hand side sums the outbound shipments over all the warehouse-to-customer lanes that originate from this warehouse. By setting these two sums to be equal to each other, we insist that warehouses can neither consume nor produce goods, but merely act as conduits to enable more efficient shipping options.

JADE Case Study Continued...

Let's get back to the new VP of supply chain and his need to quickly show supply chain savings for JADE. He had narrowed his ideas down to two main areas he felt had the potential for the largest savings for JADE. His first idea dealt with the optimization of the production network by determining the optimal product mix to produce at each plant location. The second was the optimization of warehouse locations and the customer service territory associated with each.

When considering which initiative to tackle first, he decided that changes to the manufacturing network would require more effort and take significantly more time to implement. He had found that each of their existing plants was very good within their current capabilities, and there was a level of risk associated with changing this product mix. He also presumed that there would be a lot of internal resistance to these types of changes. If he wanted to convince management of his ability to save JADE money, he concluded he should show results for the parts of the supply chain he could make changes to within a reasonable amount of time while causing the least amount of internal resistance possible. Therefore, his first objective would be to kick off a warehouse location study.

The analysis would start by reviewing differing numbers and locations of warehouses to service the JADE customers.

Transportation costs were by far the dominant cost in the warehouse network, so the VP immediately decided that the objective of his study would concentrate on minimizing that specific cost area only. All other warehouse costs were assumed to be negligible to the study. Based on what we previously learned about JADE's network, we know that the trade-off that the VP's study will be making is between the inbound cost and the slightly more expensive cost to transport product outbound to the customers.

The VP turned to a commercial network design application to facilitate his analysis. After loading all his data into the application, his first scenario was set up to select the best two locations, which he would then directly compare to the current two-location network being utilized by JADE.

Figure 9.6 shows the resultant optimal two-warehouse network. You can see that the model elects to move their original Phoenix warehouse servicing the West Coast only a slight distance to Las Vegas. The much more dramatic recommendation by this model, however, was to move the East Coast warehouse from New York to Columbus, Ohio. The VP was immediately able to understand this recommendation. Just by examining the map output, you can see that this new warehouse location is both closer to customer points and closer to the other plants.

Using these two new warehouse locations, the total transportation cost decreased to $196 million (from our original cost of $254 million). The cost to get product from plants to warehouses went from $133 million to $94 million, while the cost from warehouses to customers went from $121 million to $103 million.

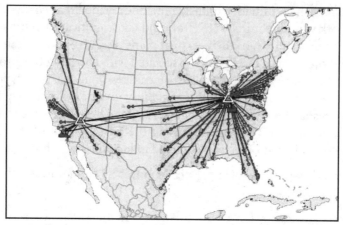

Figure 9.6   JADE Optimal Two-Warehouse Network

This solution also shows a slight improvement in the customer service JADE will be able to offer. The percentage of demand within 400 miles increased from 32% to 37%. Not a dramatic improvement, but the VP realizes that using only two warehouses will always greatly limit service-level capability across the entire country. To test for even further transportation savings and improved service, he decided to run additional scenarios allowing more warehouse locations to be added to the network.

His analysis of the best three, four, and five locations resulted in the warehouse selections shown in Figure 9.7.

Best 3 Selects:
Baltimore, Louisville, and
Las Vegas

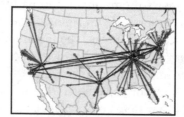

Best 4 Selects:
Baltimore, Louisville,
Las Vegas, and Dallas

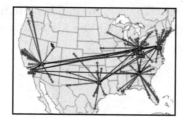

Best 5 Selects:
Baltimore, Columbus, Atlanta,
Las Vegas, and Dallas

Figure 9.7   JADE Best Three-, Four-, and Five-Warehouse Solutions

As a result of these scenarios, the VP quickly prepared the table shown in Figure 9.8 comparing the costs within each.

	Base Case	Best 2	Best 3	Best 4	Best 5
TOTAL COST (In Millions)	$254 MM	$194 MM	$182 MM	$176 MM	$169 MM
Plant to Warehouse Cost	$133 MM	$94 MM	$102 MM	$107 MM	$109 MM
Warehouse to Customer Cost	$121 MM	$100 MM	$80 MM	$69 MM	$60 MM
% Service within 400 Miles	32%	37%	63%	71%	72%

Figure 9.8   Best Warehouses Scenario Cost Comparison

His first insight is in regard to the significant savings by just changing the location of warehouses. Simply moving the two warehouses to more optimal locations reduces transportation costs significantly. Adding more warehouses then only continues to reduce the cost. It is quite clear that JADE needs to change the structure of their supply chain.

However, before we jump directly to the VP's final answer, let's review some of the deep insight into JADE's supply chain problem that this small set of data actually provides.

First, although we modeled only transportation costs, we gained some nice insight into the warehouse fixed cost as well. As we go from the best three warehouses to the best four, we see savings of $6 million. Therefore, we can conclude that if a new warehouse costs more than the $6 million, it would not make sense to open the fourth warehouse. Or, alternatively, if we cannot recoup the cost of moving into a fourth site with $6 million in annual savings, this move is not worth it. So even though the model did not capture these costs, we have enough information from the model to make decisions based on these factors.

Second, note that the rate of savings decreases with each warehouse added to the network. Simply moving to the optimal two locations, we save $60 million. By then adding a third warehouse, we see savings of an additional $12 million. Continuing on to add the fourth and fifth locations, we see that savings are only $6 million and $7 million. Based on this data, we can quickly conclude that the incremental value of additional warehouses decreases.

Third, the actual locations of the warehouses change within each solution. Sometimes, we get lucky and the optimal site locations build on each other from solution to solution. However, we cannot necessarily assume that implementing the best three locations now can be converted to the best four location solution later with the simple addition of one warehouse. If we want to eventually end up with four warehouses, we have two choices:

1. Pick the best three and add a best fourth warehouse later based on the original three locations as fixed within the model. This gets you a good solution for the best three, and if you never implement the fourth, you won't have regrets.

2. Pick the best three out of the best four locations. If you are sure you will soon implement the fourth warehouse, this choice ensures that your fourth warehouse choice will be a good one.

Of course, as with all analysis, it is good to test both cases and see whether the cost difference is significant. The benefit in modeling is that these tests are free. If the cost differences are trivial, there is no reason for debate about the risk of each of the previous options.

Our fourth insight is that the savings of additional warehouses correlates directly with the savings in the warehouse-to-customer shipment costs. Because of the relative cost difference between plant-to-warehouse ($0.07 per ton-mile) and warehouse-to-customer ($0.12 per ton-mile) transport costs, it is always better to have more warehouses and locate them close to customers. Additional warehouses drive down the warehouse-to-customer costs, but you do pay for this with additional ton-miles going from the plant to the warehouse.

Fifth, we also see an increase in percentage of customers within 400 miles as the number of warehouses within the network increases. The model has no set parameters in regard to this data; however, the information is virtually free within our analysis. As determined previously, transportation costs decrease as we get closer to customers; therefore, improvement in service levels (proximity to demand) is directly correlated. This natural correlation among scenarios brings us to another good lesson, however: There may be an objective that is important to you (demand within 400 miles), but if this objective is not included in the optimization, there is no guarantee that the correlation will remain true across all scenarios.

Based on everything we learned previously, we should always remember to be careful about the assumptions we make. Assumptions are always made in these models. If the model did not include assumptions, it would likely be as complicated and costly as the real world and do us no good for the purpose of strategic analysis and efficient decision making. A key assumption in this model, for example, is the cost per ton-mile for our various shipments. Let's test the logic of this assumption and see how it holds up.

The cost for warehouse-to-customer transit was developed by analyzing the current delivery costs to each customer. Presumably, each customer receives products from only a single warehouse, so this means that the order size and therefore shipment size will not change. In addition, as we add warehouses, logically this will lead to more short-distance hauls. Therefore, the minimum charge we have in the model helps protect us from assuming artificially low costs due to short-distance hauls. Based on these considerations, we can have confidence in these numbers.

The plant-to-warehouse assumptions may need more questioning, however. The validity of these rates depends on filling up the rail cars and trucks running on these lanes. If each plant is shipping to too many warehouses, however, it becomes more difficult to fill railcars and trucks. In this case, we can further test the ability to fill railcars and trucks by analyzing the annual flow of product from each plant to each warehouse. For instance, a total of 13 million pounds of inbound product from plant to warehouses translates into 250,000 pounds per week, which still should mean we can easily fill up our railcars and trucks with up to five total warehouse locations.

13,000,000 / 52 weeks = 250,000 lbs moving per week

250,000 lbs transferred to up to 5 warehouses = Average of 50,000 lbs to each warehouse

50,000 lbs > 45,000 lbs capacity of rail and truck transit modes

After reviewing the warehouse location scenarios and the amount of insight the VP of supply chain is equipped with, he is now in an enviable position. There are significant savings opportunities. He knows the major trade-offs, what information he gets free, and the limits of his current assumptions. Even under these circumstances, however, he still feels the need to run more scenarios to finalize his decision. You can assist him with these scenarios by completing the end-of-chapter questions.

Plant Locations Considering the Source of Raw Material

Similar to warehouse location studies, a network design study can help locate plants. In these cases we often want the location of suppliers or raw material to influence the location of the plant.

When transportation costs are important, you want to look at the same trade-off you looked at with warehouses: the difference between inbound and outbound transportation. That is, if it is expensive to ship product out of the plant relative to the cost to ship raw material into the plant, then the plant locations will likely be pulled closer to the customers.

One interesting difference, however, relates to our previous discussion around a plant receiving greater levels of inbound product compared to finished-good units being shipped out. In the warehouse examples already examined, we assumed that for each ton of product that came into a warehouse, the same ton was eventually shipped to the customer. For a plant, though, you may bring in two tons of raw material for every ton of finished good you ship out. In this case, you are still interested in the relative cost difference in inbound versus outbound transportation costs but now also with the extra influence of total tons being shipped.

Linking Locations Together for More Than Three Echelons

This three-echelon modeling logic also applies to networks containing more than three echelons. For example, let's consider a business with the need for a hub-and-spoke type of warehouse structure.

Distributors often specialize in selling products sourced from thousands of small vendors. These firms add value to the market by consolidating shipments from all these vendors. The ability of the distributor to source and distribute this wide array of products as a single entity also simplifies the customer's order process by allowing for the placement of just a single order for all needed items. The most expensive parts of a distributor's supply chain are, most often, the collection of all the products from the numerous vendors and then making the final outbound delivery to the customer.

Because of this costly inbound as well as outbound transport, it often makes sense for distributors to locate multiple levels of warehouses. As shown in Figure 9.9, we see that the first set of warehouses collects products from the vendors while the second set is responsible for the final delivery to the customers. In the middle, the company can efficiently move product in full truckloads between warehouse sets.

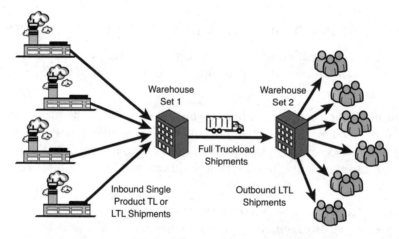

Figure 9.9   Hub-and-Spoke Network Depiction

When locating the first set of warehouses, there are two common strategies. The first strategy locates either a single or small number of central warehouses. Locating just a few centrally works well when the vendors have enough volume to ship in full truckload to one or a few points. The second strategy locates numerous warehouses close to vendors and those vendors then ship short LTL shipments into the closest warehouse. This strategy works well when the vendors cannot fill a full truck even when shipping to just one location.

When locating the second set of regional distribution centers, common practice is to select numerous locations close to customers. As we have seen previously, the company's ability to transport product in full truckloads into these warehouses allows for the concentration on better service and shorter outbound LTL shipment distances to customer points.

For some companies, however, two echelons of warehouses may still not be an optimal structure. Think about a major retailer that sells product online. Consumer locations become people's houses, and orders may be as large as an appliance or as small as a single DVD or a package of socks. Based on what we have previously learned about transportation, these small shipments can be expensive to deliver to the home, even with the existence of regional distribution centers. For larger items (like appliances), companies will commonly use small trucks making multiple stops in a small geographic area (like a town or part of a metro area). These multistop routes require a local presence to facilitate their local delivery process, however. For the smaller items, these companies may use small-parcel delivery services. But even so, the objective would be to keep the product on ground and in full trucks as long as possible to control costs. So in both previous examples we see the need for the location of a third and final set of local warehouses. These locations are meant to receive product from regional warehouses and ease the high cost of small shipments deep into their markets.

Because it can be expensive to operate a chain of fully stocked local warehouses, many firms, in practice, treat these warehouses as cross-docks. It usually works something like this:

1. Orders come in throughout the day for deliveries needed the next day.

2. After some cutoff time, the regional warehouse starts picking orders for the local warehouses. These orders are loaded onto a full truck.

3. Overnight, the truck drives to the local warehouse.

4. First thing in the morning, the product is moved from the regional warehouse truck onto the delivery vehicles.

This approach has the benefit of moving the product as far as possible on a full truck.

When firms ship small-parcel and are charged per package, it is often beneficial to employ a strategy called "zone-skipping," which is analogous to the previous example. The "zones" we are skipping are the zones set up by the small-parcel carriers (UPS, FedEx, DHL, and so on). The more zones we have to ship through, the more expensive the cost per package. Now, if we have enough volume, we can fill up our trucks with the packages for a local market and drive these trucks straight to the parcel carrier's local sort facility. Instead of our delivery trucks, we are giving the product to the parcel carrier's delivery trucks.

The previous hub-and-spoke networks had fixed supply points. What happens when companies need to determine optimal plant locations at the same time? In this case, we quickly find ourselves with more than three echelons to include in our analysis. When we locate our plants, we want to consider the location of the raw material sources as well as the location of the warehouses. And when we locate the warehouses, we want to consider the location of the plants and the location of the customers. So, in this model, we would be locating both the plants and the warehouses simultaneously. But as you are becoming a skilled modeler, you must always evaluate the complexity of the models you are proposing and the complexity of the output it may provide.

It is common to encounter a supply chain analyst having more difficulty in analyzing and interpreting the results than in setting up the model. When a model returns many different decisions at the same time, it can be difficult to explain what happened, spot potential problems, and determine the appropriate alternative scenarios. These types of difficulties can easily derail an entire project, and no CEO is going to make changes to the network if those changes cannot be fully explained and backed up by clear analysis.

One of our main objectives in this book is to break down supply chain network analysis into simple models you can use as building blocks to facilitate learning and allow you to efficiently analyze your complex problems. By using a series of simple models, you may be able to more effectively analyze your model without losing accuracy.

In the hub-and-spoke example, the problem naturally splits into two problems. The first is locating the central warehouse close to the suppliers. As previously discussed, thanks to the ability to ship relatively low-cost full truckloads from the central to regional warehouses, we can locate the central warehouses independent of customer locations. In parallel, the location of the regional warehouses does not depend on the central warehouses either. Instead, it depends solely on the location of the customers. By separating the models, we have two models that we can easily analyze and understand.

Even expert modelers know that the key to good modeling is in keeping it as simple as possible while still producing the level of detail required for explaining the results and eventually implementing the suggestions. By starting simple, you will also benefit from a full understanding of each modeling capability before moving on to include more complexity. If you ever find yourself getting confused, your best bet is to return to a single subset or the simplest portion of the analysis and work your way through from there. Chapter 12, "The Art of Modeling," will further build on this concept as well.

Lessons Learned from Three-Echelon Supply Chain Modeling

When modeling three echelons at the same time, we were introduced to a new trade-off between the inbound and outbound costs. The inbound costs to a facility will tend to pull the location close to the source of products. The outbound costs from a facility will tend to pull the location close the customers. The relative difference between the inbound and outbound costs will determine which side has more pull (the sources or the customers).

Once we understand the concept of three echelon supply chains, we can naturally extend our models to include any number of echelons. We just need to watch out for the extra complexity that we add.

End-of-Chapter Questions

You can find the JADE model as well as detailed instructions for reviewing and running additional scenarios for the JADE Case Study within the JADE Case Study Exercise.zip file on the book Web site. Follow the directions within the JADE Intro Case.ppt file to open the model and review all input and output files to start. Then use this basis as well as additional scenario-building instructions to answer the following questions:

1. Based on the results seen previously in this chapter, as well as your running of these scenarios on your own through the model and PowerPoint referenced previously, review the new locations selected as the model solves for scenarios with two to five optimal warehouses and then explain:

   a. How much impact did the change from Phoenix to Las Vegas have? Why?

   b. What is the rate of change in the savings as each additional warehouse is added?

2. In the *model*, why would you expect the costs to continue to decrease as you add warehouses? In *reality*, would you expect this to hold up? Why or why not?

3. In the five-warehouse solution, even though the Columbus warehouse is very close to the plants, it still chooses to serve customers from additional warehouses. Why? In other words, why not save on inbound transportation and serve most customers from Columbus?

4. What if we wanted to implement five warehouses in the future, but wanted only four now. What is the cost of the best four, plus adding one more, versus the cost of picking the best four out of the best five? Which solution would you recommend?

5. Even though we don't have the cost of opening a warehouse in the model, what information does the model give us in terms of what this cost needs to be for these solutions to be effective?

6. What are the optimal three warehouses and network total cost if all product comes from China (enter through the Port of Long Beach)?

7. What would have been the cost if they moved all production to China, but still chose our previous optimal three-warehouse solution (dual sourcing product)?

8. What is the value if all plants could make all four products? How much better of a solution could you derive?

9. Assume that the CEO is not sure whether all product will come from China but is sure that all plants will not make all products. What is your recommendation to the CEO for the location of the best three warehouses? Prepare three or four slides and be ready to present your result. You have to make a recommendation, but you are free to show other alternatives and discuss their merits. If there is missing data or information, feel free to use whatever assumptions you want, but make sure that those assumptions are as realistic as possible.

10

ADDING MULTIPLE PRODUCTS AND MULTISITE PRODUCTION SOURCING

In the previous chapters, we have focused on various aspects of network design, including center of gravity models, service levels, and transportation cost modeling. All of these are key elements to understanding network modeling. However, we have yet to focus on differentiating between various types of products as they move through the supply chain. In this chapter, we will look at adding multiple products to the analysis and understanding its importance as it relates to network design.

Why Model Products?

Firms typically have thousands if not tens of thousands of products or SKUs that they manage and distribute through their supply chain. As we have seen in previous chapters, it is often common practice to use overall product weight to represent their movement within network models. We are in essence trying to estimate and represent the transportation costs associated with moving these products to meet customer demand. And as we have learned, transportation costs are heavily impacted by weight. Therefore, for modeling purposes, the appropriate costs will often be captured if we ensure that the total weight is accurately represented in the model. This also assumes that all products are rated uniformly using the same transportation rates. If that holds true, it is safe to assume that our results would be equivalent when modeling this weight moving in the form of 1 generic product or 100 defined products.

However, in this chapter we will learn that there are several cases where modeling a generic product may not be appropriate—in fact, when doing so, our model may yield incorrect results. These cases include the following:

- When our overall pool of products includes large variations in storage and logistics characteristics

- When some product types require specific customer service levels and therefore require specific transport modes

- When products come from different source locations

Let's review each of these cases in further detail.

Variations in Logistics Characteristics

As we have discussed before, firms typically carry and distribute thousands, if not tens of thousands, of SKUs across the supply chain. It may be likely that these SKUs fall within different product-family designations, each with its own distinct characteristics, therefore requiring the modeling of separate products to accommodate the logistical differences as well.

Let's take the example of a retailer. The SKUs carried by the retailer may include dry packaged goods (such as boxes of cereals, cans of soup), fresh produce items (apples, bananas, lettuce and so on), and frozen products (such as frozen dinners, frozen vegetables, and ice cream).

Each of these product families may be distributed in similar types of packaging (cartons), and they may even have roughly the same product density. However, the storage and transportation requirements are still very different for each group of items. The dry packaged goods can be transported in regular dry van trucks and stored at normal room temperature in the warehouses and stores. Fresh produce and refrigerated items, however, require specialized trailers with onboard refrigeration units that maintain the right temperature to prevent spoilage during transport. The transportation rates for reefer trailers (trailers with onboard refrigeration units) are typically 10% to 30% higher than those of standard dry vans. Bearing this in mind, we will need to ensure that we separate this volume of products moving on lanes so that they can be modeled with the correct transit cost using these higher transportation rates.

The same parameters also apply when it comes to storage requirements in our models. It is much more expensive to set up and run a warehouse with temperature-controlled chambers for refrigerated and frozen products. As a result, it is common for firms to use a large network of warehouses for dry packaged products, but storing refrigerated or frozen products in one or two warehouses only, given the higher capital requirements needed to support this capability.

The same rules are true for products that are considered hazardous materials or "hazmat." These products have special storage requirements in the warehouse, as well as special handling constraints for transportation, thereby incurring higher costs. To address these constraints effectively, we again must ensure that we separate their activity within models by representing these as separate products.

Products with Differing Service-Level Requirements

When products are modeled at an aggregate level for each ship-to location, the analysis would essentially look at treating the shipment of all demand to customers in the same way. However, the shipment methods and patterns may actually be very different when analyzed in conjunction with demand by product families.

As you can see in the example in Figure 10.1, there is a significant difference in demand distribution between the two product families. This information can help in better analyzing and developing an optimal distribution strategy. For example, it is common for firms to develop different distribution strategies for distinctly different types of product families. SKUs that are considered critical parts may require expedited service such as same-day shipping or next-day delivery. The rest of the product portfolio may be delivered through normal shipping methods. In this case, it would be important to break out and model these products separately so that the appropriate service level constraints can be applied and the right network strategy can be developed.

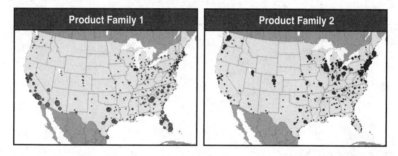

Figure 10.1 Example Showing Differences in Demand Demographics Between Product Families

Leading companies tend to treat their supply chain as a collection of multiple supply chains for the various product families. The best overall modeling strategy is to develop the right network and distribution strategy for key product segments, while making sure that this does not create a suboptimal network at an aggregate level.

Product Sourcing

The source for the product is probably the most overlooked and important reason to include multiple product groups in a model. In the model, we want to make sure products come from the correct locations. Including product sourcing as part of our product groups, we ensure that this happens.

Including product sourcing information in our product groups allows us to specify which products come from which plant and which products are allowed to come from multiple plants (and we can specify which of the multiple plants). This ensures that our model correctly models the flow of product from the source to the final destination.

Adding Products to the Model—Mathematical Formulation

There are many ways to formulate a model with multiple product families. The simplest enhancement to this, however, would be to add product indexing $k \in K$ to various components of the supply chain. This model is quite sophisticated so the following detailed analysis should give you further intuition.

$$\text{Minimize} \quad \sum_{i \in I} \sum_{j \in J} \sum_{k \in K} (transWC_{i,j,k} + whVar_{i,k}) d_{j,k} Y_{i,j,k} +$$

$$\sum_{l \in L} \sum_{i \in I} \sum_{k \in K} (transPW_{l,i,k} + pVar_{l,k}) Z_{l,i,k} +$$

$$\sum_{i \in I} \sum_{w \in W} whFix_{i,w} X_{i,w}$$

Subject to:

$$(1) \sum_{i \in I} Y_{i,j,k} = 1; \forall j \in J, \forall k \in K$$

$$(2) \sum_{i \in I} \sum_{w \in W} X_{i,w} \geq P_{\min}$$

$$(3) \sum_{i \in I} \sum_{w \in W} X_{i,w} \leq P_{\max}$$

$$(4) \sum_{w \in W} X_{i,w} \leq 1 \forall i \in I$$

$$(5) \sum_{j \in J} \sum_{k \in K} vol_{i,j,k} Y_{i,j,k} \leq \sum_{w \in W} whCap_{i,w} X_{i,w}; \forall i \in I$$

$$(6) \sum_{l \in L} Z_{l,i,k} = \sum_{j \in J} d_{j,k} Y_{i,j,k} \forall i \in I, \forall k \in K$$

$$(7) \sum_{i \in I} Z_{l,i,k} \leq pCap_{l,k}; \forall l \in L, \forall k \in K$$

$$(8) Y_{i,j,k} \leq \sum_{w \in W} X_{i,w}; \forall i \in I, \forall j \in J, \forall k \in K$$

$$(9) Y_{i,j,k} \in \{0,1\}; \forall i \in I, \forall j \in J, \forall k \in K$$

$$(10) X_{i,w} \in \{0,1\}; \forall i \in I, \forall w \in W$$

$$(11) Z_{l,i,k} \geq 0; \forall l \in L, \forall i \in I, \forall k \in K$$

The objective function here consists of three components—the shipping from plant to warehouse, the shipping from warehouse to customer, and the warehouse fixed costs. (Note that in this model, we assume, for the sake of expediency, that plants cannot ship directly to customers. This useful extension, like many others, is not difficult to incorporate.)

The summation

$$\sum_{i \in I} \sum_{j \in J} \sum_{k \in K} (transWC_{i,j,k} + whVar_{i,k}) d_{j,k} Y_{i,j,k}$$

captures the cost of shipping from warehouses to customers. We have many terms that are similar to those we have discussed previously, so it shouldn't be very difficult to understand how this summation works. The set K represents our set of products (or, if product aggregation is used, our set of product families). We use the variable index k to represent a specific product (or product family). Thus, we use the term $d_{j,k}$ to represent the demand for product k at customer j. We similarly capture the cost to process product k at warehouse i with the term $whVar_{i,k}$. The series of summations $\sum_{i \in I} \sum_{j \in J} \sum_{k \in K}$ simply indicate that we are summing the warehouse-to-customer shipping and processing costs for every warehouse, customer, and product triplet.

The summation

$$\sum_{l \in L} \sum_{i \in I} \sum_{k \in K} (transPW_{l,i,k} + pVar_{l,k}) Z_{l,i,k}$$

is similarly used to capture the cost of shipping from plants to warehouses. The term $transPW_{l,i,k}$ is used to represent the cost of shipping one unit of product k from plant l to warehouse i. The term $pVar_{l,k}$ is used to represent the cost of producing one unit of product k at plant l. The variable $Z_{l,i,k}$ is used to represent the amount of goods of product k that are shipped from plant l to warehouse i. The summation $\sum_{l \in L} \sum_{i \in I} \sum_{k \in K}$ indicates that we are performing this sum for every plant, warehouse, and product.

Note that Z is neither an integral nor a binary variable. That is to say, this variable is *continuous*. It is allowed to take on any value, such as 0.5, 100.3, or 1212.3. However, if we use one of the more common types of MIP solver engine, and all of our capacity and demand information is integral, then we know that any optimal solution will have an integral result for the variable. In practice, it is recommended that a user encourages integrality through this method, or, even better, simply rounds the results for shipping variables like this. Although there are more explicit methods of enforcing integrality on the per-product shipping variables, they almost always result in a significant increase in runtime.

The family of constraints

$$(1) \sum_{i \in I} Y_{i,j,k} = 1; \forall j \in J, \forall k \in K$$

is very similar to the one we discussed previously. We are simply enforcing that a warehouse be selected as the service provider for each demand record. Note that we are prohibiting product splitting, but we are not enforcing single sourcing. That is to say, we will allow a customer to receive different goods from different warehouses, but we won't allow a customer to receive the same product from more than one warehouse. It is possible to alter our model to enforce any of the logical combinations for these settings. That is, a customer might require single sourcing (and thus implicitly prohibit product splitting), allow multiple sources but not product splitting (as we do here), or allow product splitting (and thus not enforce single sourcing).

The constraint families (2) and (3) are not new. They simply require the model to select at least P_{min} warehouses to open, while ensuring that no more than P_{max} are chosen.

Constraint (4) has also been discussed before. It simply says that if a warehouse is opened, the solver must select one of the warehouse options from set W to build at this site.

Equation (5), while similar to a constraint family developed before, is worth closer examination.

$$(5) \sum_{j \in J} \sum_{k \in K} vol_{i,j,k} Y_{i,j,k} \leq \sum_{w \in W} whCap_{i,w} X_{i,w}; \forall i \in$$

In particular, we can draw attention to a few things. First and foremost, we note that the left-hand summation is over all the demand points (i.e., all the customer/product pairs). For a given warehouse, the left-hand side of this equation is the total volume of the goods being stored in inventory. The right hand side represents the size of the warehouse. The constraint family ensures that the warehouse will not hold more inventory than its size will allow.

Second, it is worth pointing out that the term, $vol_{i,j,k}$ respresents the amount of space consumed by one unit of product k stored at warehouse j bound for customer i. It takes into account the following distinct quantities:

- The total demand of product k required by customer j. Although we have a distinct term $d_{j,k}$ that captures this value precisely, we are blending this quantity into the value $vol_{i,j,k}$ for simplicity's sake.

- The physical dimensions of product k. For warehouse capacities, this quantity might be measured with something as simple as cubic volume or pallet positions, or it might be measured in a more sophisticated way that captures the efficiency of floor packing.

- The "inventory speed" of the warehouse. That is to say, a warehouse that frequently turns its inventory has a higher effective capacity than one whose goods linger for a long period. Because "inventory speed" (more formally known as inventory turnover ratio) varies from one product to the next, we need to incorporate this term into the *vol* term, instead of the *whCap* term.

This is just a cursory summary of how aggregate warehouse capacity is measured in a strategic network design model. We could write an entire chapter on this subject alone. For our purposes, it is sufficient to give you a sense of the complexity involved just in this one constraint family.

The constraint family

$$(6) \sum_{l \in L} Z_{l,i,k} = \sum_{j \in J} d_{j,k} Y_{i,j,k} \forall i \in I, \forall k \in K$$

captures a well-known type of constraint called "conservation of flow." Essentially, we are saying that the warehouse can neither create goods nor consume goods. Instead, goods enter the warehouse and leave the warehouse in equal quantities.

The left-hand side of the equation, $\sum_{l \in L} Z_{l,i,k}$, captures the total demand of product k inbound to warehouse i. The right-hand side, $\sum_{j \in J} d_{j,k} Y_{i,j,k}$, captures the total outbound shipments of the same product. By forcing these two sums to be equal for every warehouse/product pair, we ensure that a warehouse is allowed to neither create goods nor consume them.

The constraint family

$$(7) \sum_{i \in I} Z_{l,i,k} \leq pCap_{l,k}; \forall l \in L, \forall k \in K$$

ensures that our plants are not used beyond their capacity. We are applying a capacity at the plant-product level. That is to say, we are ensuring that plant l cannot produce more than $pCap_{l,k}$ units of product k. In industrial formulations, this constraint is often complemented by a plant-specific constraint that restricts some form of total production over all products.

Constraint families (8), (9), and (10) are very similar to those we developed previously. Constraint (8) has been modified slightly to ensure that a demand point cannot be assigned to an unopened warehouse (as opposed to the previous formulation, which ensured that a customer could not be assigned to an unopened warehouse).

Constraint family

$$(11) Z_{l,i,k} \geq 0; \forall l \in L, \forall i \in I, \forall k \in K$$

simply insists that the plant-to-warehouse shipping variables be nonnegative. Note that, unlike constraint families (9) and (10), we do not insist on true integrality here.

We now see the importance of adding products to a model, as well as how the addition of products essentially adds another entity with its own set of unique variables. This allows for modeling additional constraints such as restrictions on which products can flow through each warehouse, as well as setting the maximum number of warehouses that a given product family may be stored within.

To further our understanding of the inclusion of products in a network model and evaluate the implications of the solution, let's go through a case study.

Case Study—Value Grocers, Grocery Retailer

Value Grocers is a mid-size grocery retailer headquartered in Chicago, Illinois. The retailer operates a network of 120 stores across the Midwest portion of the U.S. The retailer specializes primarily in an array of grocery items from frozen vegetables to juices and beverages. The company was founded in Chicago with one store operating out of a Lincoln Park neighborhood on the north side of Chicago. Based on the success of their first store, the founder decided to expand into other parts of the city. The business continued to grow aggressively over the next decade, and the Value Grocers' store network expanded throughout the Midwest region, including store locations in Wisconsin, Minnesota, Iowa, Michigan, Ohio, Indiana, and Kentucky. The map in Figure 10.2 shows the current store network.

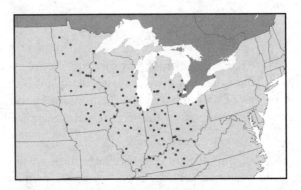

Figure 10.2   Value Grocers Store Locations

The company currently operates a single large warehouse in Chicago, which was set up originally when the company consisted of a small network of stores within the Chicago area. Despite geographic expansion across seven states over a decade, the company continues to serve all stores from this Chicago warehouse.

With the increase in fuel prices and overall transportation rates, the management team has become concerned about the growing logistics spend. With increased consumer focus on product freshness, the management team was also interested in increasing service levels so the stores could be replenished faster. Keeping these objectives in mind, the management team decided to initiate a network analysis to evaluate using additional warehouses to serve their stores.

For simplicity, we will use a representative set of products that are stocked and distributed by Value Grocers:

- Frozen products—for example, frozen beans, frozen corn
- Bottled juices—for example, pineapple juice, orange juice, apple juice
- Canned/bottled beverages—for example, soda and diet soda
- Packaged goods—for example, cereals, bread, nutrition bars

At first glance, it might be tempting to consider all of these items as one "grocery" product and focus on the total weight shipped from the Chicago warehouse to each Value Grocers store. However, these items have largely differing storage and logistics characteristics that require us to model several product groups as opposed to one flow of goods. We should first note that the frozen products and chilled juices require a temperature-controlled environment during storage and transportation. Therefore, as previously discussed, we will need to ensure that we reflect the higher transportation rates associated with the reefer (refrigerated) trailers that will be required to transport them.

Understanding the Impact of Product Dimensions on Transportation

In addition to the temperature requirement differences, let's review the products' dimensional characteristics as well. The table in Figure 10.3 shows the unit carton weight by product family. Looking at this information, we see that juice and beverage items are heavier, most likely due to their high water content, whereas cereal boxes and baked goods are relatively lighter. Why is this important to us? Let's do a small side analysis to get a better understanding.

Name	Weight (lbs.)
Frozen vegetables	20.000
Fresh juices	30.000
Canned beverages	36.000
Cereals and bars	12.000

Figure 10.3   Weight (Lbs) per Carton for Key Product Families

Let's first assume that all product groups are packaged in similar sized cartons—each with a volume of two cubic feet. If we were to fill up an entire truck with cartons of any one product family, how many cartons could be loaded based on their dimensions?

All trailers have a maximum capacity in terms of weight and cube (volume), and the true capacity is based on whichever limit is reached first. Let us assume that the maximum weight limit for the trailer (53' trailer) in this example is 40,000 pounds, and the maximum capacity based on volume is 4,000 cubic feet.

The results of this initial analysis are shown in Figure 10.4. For fresh juices, we can load a maximum of 1,333 cartons based on weight and a maximum of 2,000 cartons based on cube. This means that we will hit our weight limit before we hit our volume limit and therefore can load a maximum of only 1,333 cartons within each trailer for distribution. We can also refer to this as "weighing out"—a common industry term for heavy products that are constrained in transport capacity by weight limits.

Product Type	Carton Unit Wt (lbs)	Carton Vol (cuft)	Truckload - Max Weight (lbs)	Max # of Cartons Based on Weight	Truckload - Max Volume (cuft)	Max # of Cartons Based on Volume
Frozen Vegetables	20	2.0	40,000	2,000	4,000	2,000
Fresh Juices	30	2.0	40,000	1,333	4,000	2,000
Canned Beverages	36	2.0	40,000	1,111	4,000	2,000
Cereals and Baked Goods	12	2.0	40,000	3,333	4,000	2,000

Figure 10.4 Product Characteristics Versus Truckload Capacity

On the other hand, cereal and baked-goods products will hit the max capacity based on volume (i.e., 2,000 cartons versus 3,333 cartons). These types of product are considered to be "cubing out" because they are lighter in density.

So how do these cube and weight differences impact our network model? As we know by now, transportation costs by lane is a key input into our network models. The weight and cube characteristics just reviewed will affect the amount of product that can be shipped for each iteration of truckload cost and therefore will also impact the transportation cost per unit calculated and used by the model. Figure 10.5 shows the calculation of transportation cost per unit for our data.

Product Type	Carton Unit Wt (lbs)	Carton Vol (cuft)	Truckload - Max Weight (lbs)	Max # of Cartons Based on Weight	Truckload - Max Volume (cuft)	Max # of Cartons Based on Volume	Transportation Cost per Unit (TL) Cost = $1500
Frozen Vegetables	20	2.0	40,000	2,000	4,000	2,000	$0.75
Fresh Juices	30	2.0	40,000	1,333	4,000	2,000	$1.13
Canned Beverages	36	2.0	40,000	1,111	4,000	2,000	$1.35
Cereals and Baked Goods	12	2.0	40,000	3,333	4,000	2,000	$0.75

Figure 10.5 Transportation Cost per Unit by Product

For this example, if the truckload cost for a given lane was $1,500, the transportation cost per unit for canned beverages could be calculated as $1.35/carton ($1,500 / min [1111, 2000]), and cereals and baked goods would cost $0.75/carton. This leads us to easily conclude that it would cost almost 80% more per unit to ship a carton of canned beverages versus a carton of cereal products.

The previous analysis helps us, as modelers, in a couple of ways. First, it allows us to understand how to accurately calculate and represent transportation costs for these products. Second, this difference in unit transportation costs we discovered will directly impact several cost trade-offs in our analysis, and thereby impact where and how many warehouses these products are served from. We will revisit this topic in more detail later in this section.

To summarize, we can now firmly conclude that it is important to our model that we differentiate between several different product families in this analysis—not only due to the products' varying temperature requirements but also because of their widely varying per-carton product weights. For simplicity, we will model four product families in this model—each product is actually an aggregated representation of hundreds of SKUs within the respective family. We will cover aggregation strategies in more detail in subsequent chapters.

Distribution Network Analysis Based on Outbound Flow

Now that we have defined our products, we can estimate demand by product and store and include this in our model. The map and chart in Figure 10.6 show a demand map with each store (represented as circles) sized by its relative product demand, and a bar graph showing total demand by product. The map highlights that Illinois, especially around metropolitan Chicago, still represents the high-demand areas—not surprising given the history of the company in this area. Stores in large cities such as Minneapolis, Detroit, and Cleveland also represent larger demand points compared to those in the non-urban areas. This is likely directly attributed to the relative differences in population density.

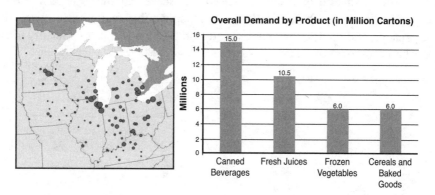

Figure 10.6   Map of Demand by Store and Demand by Product

Next, we will add a list of potential warehouses to evaluate as part of this network. The transportation in this model will consist of two types of carriers—one representing the use of regular dry-van trailers, and one representing temperature-controlled reefer trailers. Both carriers represent full-truckload transport with an associated cost-per-mile rate and minimum charge. Figure 10.7 shows the higher per-mile cost associated with the reefer trailers. Both modeled carriers' transportation capacities will be set as a maximum of 40,000 pounds or 4,000 cubic feet and the model will apply the capacity limit that is first reached by product family.

Name ▲	Rated Cost ($/Mile)	Min Charge ($)	Weight (lbs.)	Volume (cu ft)
Regular Dry Van TL	2.00	300.00	40,000	4,000
Reefer TL	2.80	350.00	40,000	4,000

Figure 10.7   Transportation Rates for Delivery to Stores

We will now begin running several scenarios to test where the best one, two, three, and four warehouses would be located. Our first objective, within the best one-warehouse scenario, is to understand whether there is a better location than the current warehouse located in Chicago. Obviously, setting up a new facility would require capital investment and fixed costs. Only if the transportation savings are significant would it would make sense to relocate the existing warehouse.

By reviewing the output of each of our four scenarios (shown in Figure 10.8), we can make a few quick observations. Chicago is selected as the best single-warehouse location—this validates that their current warehouse is indeed the optimal location from which to serve all stores in a one-warehouse solution. We also notice that the model progressively selects warehouse locations based on large pockets of demand. For instance, the best two-warehouse scenario selects Madison to serve metro Chicago, Minnesota, Wisconsin, and Iowa, and Ft. Wayne is selected to serve Indiana, Michigan, Ohio, and Kentucky. The three-warehouse solution then splits the region into three distinct demand pockets, and the four-warehouse solution essentially locates a facility in each of the key urban centers of demand.

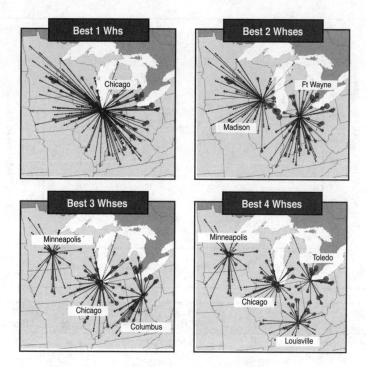

Figure 10.8   Summary of Results with Best One-, Two-, Three-, and Four-Warehouse Locations

All these observations were strictly based on what we can infer from the geography of the locations picked. Let's now take a closer look at the cost impact displayed in Figure 10.9.

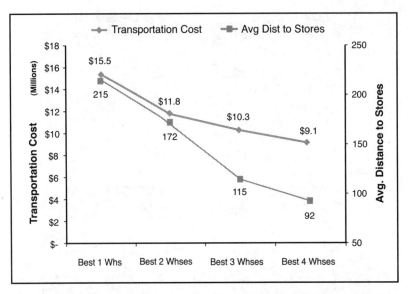

	Best 1 Whs	Best 3 Whses	Best 3 Whses	Best 4 Whses
Transportation Cost ($ Millions)	$15.50	$11.79	$10.34	$9.06
Incremental Savings	-	$3.71	$1.45	$1.28
Avg Distance to Customer (miles)	215	172	115	92
Reduction in Distance	-	43	57	23

Figure 10.9 Scenario Results Comparison

The cost-comparison table shows the biggest savings can be realized when increasing from a single optimal warehouse solution to an optimal two-warehouse solution. While subsequent scenarios continue to reduce costs as we increase the number of warehouses selected, we see that the *incremental* savings for each decreases. This reduction in marginal savings is driven by two factors. First, the overall average distance to customers will continue to reduce with each additional warehouse added to the network; however, the service level information shown in the figure clearly highlights that the *proportion* of the overall mileage saved reduces from left to right. Second, the carrier minimum charge starts to become a factor in how much savings may be realized with additional warehouse locations when the average distance to stores falls below a certain mileage band. This means that the cost of the shipment will not reduce even if an additional warehouse allows for product to be placed closer to the store than in the previous scenario.

Although we didn't include the fixed cost of each incremental warehouse within the model (we learned about the inclusion of fixed and variable costs previously in Chapter 7, "Introducing Facility Fixed and Variable Costs"), the transportation savings with the addition of each warehouse may now be easily compared against the fixed cost of each additional warehouse outside of the model. In this case, the management team made a general estimate that the fixed cost for a new warehouse would be approximately $2 million. This means that the savings found in our optimal three-warehouse solution would still outweigh the cost of the additional two facilities ($5.16 million freight savings versus $4 million warehouse costs). Within the optimal four-warehouse network, however, the transportation savings of $6.44 million are almost equivalent to the additional fixed cost for three new warehouses ($6 million). Given that the savings in distance to customer are marginal with a four-warehouse versus a three-warehouse solution (only 23 miles closer on average), it makes sense to decide that the three-warehouse network solution is the best recommendation for Value Grocers at this point. Implementing this solution translates into the addition of new warehouse facilities in Minneapolis and Columbus.

Adding Storage Restrictions for Temperature-Controlled Products

As we begin to look at the three-warehouse solution more closely, we see that all three warehouses within the network are used to store and distribute all product families to their respective stores. From a warehousing perspective, this means that both new warehouses will need to be configured with temperature-controlled rooms, which requires a significant capital outlay they hadn't fully incorporated into their average new-warehouse fixed cost estimated previously. The management team now wanted to understand the impact of storing temperature-controlled products in only one warehouse in order to minimize these additional capital requirements.

Because our modeled products are grouped by family and temperature requirements can be directly tied to the family (temperature-controlled products such as frozen vegetables and juices may be cleanly separated from our other regular, dry products), we are able to easily add this constraint in our model. The chart shown in Figure 10.10 compares the total throughput by product by warehouse for our previous optimal three-warehouse solution to the new scenario with temperature-controlled storage restrictions. This graphic clearly shows us that the Chicago warehouse is selected as the best location to store and distribute all temperature-controlled products.

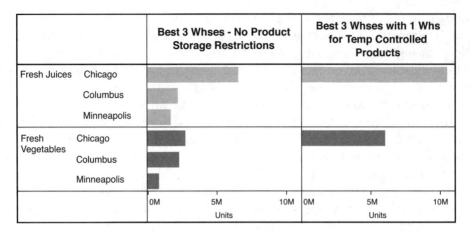

		Best 3 Whses - No Product Storage Restrictions	Best 3 Whses with 1 Whs for Temp Controlled Products
Fresh Juices	Chicago		
	Columbus		
	Minneapolis		
Fresh Vegetables	Chicago		
	Columbus		
	Minneapolis		
		0M 5M 10M Units	0M 5M 10M Units

Figure 10.10   Throughput Comparison by Warehouse for Temperature-Controlled Products

The cost comparison in Figure 10.11 shows that the solution with a single warehouse with temperature-controlled storage capabilities costs $2.42 million more than our original three-warehouse solution. This cost increase is expected because these products will be served from one warehouse instead of three, thereby increasing the average distance to stores. Obviously, this increase in transportation costs must then be compared with the savings in capital associated with equipping temperature-controlled rooms for the other two warehouse locations. If the cost to do so is more than $2.42 million, we know that this is our better solution.

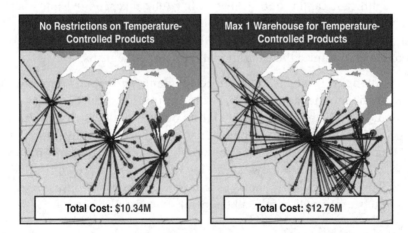

No Restrictions on Temperature-Controlled Products

Max 1 Warehouse for Temperature-Controlled Products

Total Cost: $10.34M

Total Cost: $12.76M

Figure 10.11   Maps Comparing Results of Best-Three-Warehouse Solution with and without Storage Restrictions

Based on the detailed analysis we just reviewed, the Value Grocers management team now has the necessary information to make the best decision on the optimal distribution strategy for each of its product groupings in the context of their entire network.

Addition of Product Sourcing

In the analysis we have done so far in this chapter, we looked at only the outbound distribution from warehouses to store locations. From a product perspective, however, we have yet to analyze or include where the products were sourced from and how they were shipped into these warehouses to begin with. As we discovered in the preceding chapter, introducing three-echelon supply chains, this leg of any supply chain network can have an impact on your distribution strategy.

To effectively analyze product sourcing, we need to analyze and accurately model products so that the differences in sourcing are captured appropriately. Let us continue with our Value Grocers analysis and understand how this information is incorporated into our network design study.

To illustrate this point, let's take one product category, Fresh Juices, and show how it has an impact on the solution. (In reality, each product category has an impact, and we would need to consider all of them together.) Let's take apple juice and orange juice as two specific examples. A quick analysis of inbound for these products shows that the apple juice products are being sourced from a vendor in Michigan, and orange juice products are purchased from a different vendor in Florida. In the U.S., apples are largely grown in certain key states including Washington and Michigan, whereas Florida is known for its oranges. It makes complete sense that the juice vendors, in turn, typically locate their plants close to these associated produce fields.

Value Grocers currently pays for this inbound freight from these juice vendors to its existing Chicago warehouse, so this portion of their network is important to include in our analysis. As we have learned in the previous chapters, when we look at three-echelon supply chains, the trade-off between inbound versus outbound transportation costs becomes an additional factor in determining the location of warehouses. In this case, the inbound costs from vendors in Michigan and Florida into the warehouses will be traded off with the outbound costs from the warehouses to the stores.

Before we start analyzing the impact of inbound sourcing, we will need to update the products in the model to break out the Fresh Juice product into apple juice and orange juice. Figure 10.12 shows the updated demand chart with demand for apple and orange juices, along with their product characteristics in the model.

For the purposes of simplicity and ease of analysis in this study, we will assume that all other products have local sources near each warehouse and therefore their inbound costs will be insignificant within this scenario.

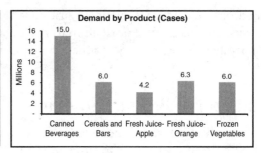

Name	Weight (lbs.)	Transportation Volume (sq. ft)
Canned Beverages	36.000	2.000
Cereals and Bars	12.000	2.000
Fresh Juice - Apple	30.000	2.000
Fresh Juice - Orange	30.000	2.000
Frozen Vegetables	20.000	2.000

Figure 10.12   Updated Products for Model and Overall Demand by Product

The apple juice vendor is located in Grand Rapids, Michigan, and the orange juice supplier is located in Ocala, Florida. Both vendors currently ship product in full truckloads into the Chicago warehouse. For this analysis, we will assume that these inbound transportation rates ($/mile) are the same as applied for outbound truckload shipments in our previous models.

Also for the purposes of simplicity and ease of analysis, we will ignore the restrictions on the number of warehouses that may store temperature-controlled products. In other words, we will compare these results with our original optimal three-warehouse scenario.

The outbound solution maps in Figure 10.13 show that the inclusion of inbound transportation costs for juice products has dramatically changed the selection of warehouse locations in the optimal three-warehouse solution. We see that the Minneapolis warehouse, originally serving the upper Midwest, has now been replaced by a warehouse slightly farther south in Milwaukee that serves both the Chicago/Illinois area and the Upper Midwest. The warehouse location serving the Ohio region has now moved north to Toledo, serving both the Ohio and the Detroit areas, and the third warehouse has been chosen in Louisville, Kentucky.

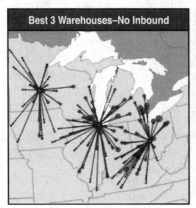
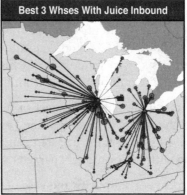

Figure 10.13   Map of Results for Best Three Warehouses—with and without Juice Inbound

What are the potential drivers of this drastic shift? Because we know that the only thing we changed from our previous solution was the inclusion of inbound transportation costs, we know this is the root cause of the changes in results. It is the result of inbound costs from Michigan and Florida that are *pulling* the warehouses both eastward and south based on the location of these vendors.

Before we look at cost comparisons between these scenarios, however, we need to run one more scenario that we can use as a fair comparison with this new scenario. Because the Chicago-Minneapolis-Columbus scenario did not include these inbound transportation costs, this scenario cannot be used as the basis of comparison with the new scenario. As a result, we will run another scenario with inbound juice transport and the original selection of Chicago, Minneapolis, and Columbus as the three-network warehouses, but allow the store assignments to be optimized. This scenario will then have the exact same cost components as our new scenario and can be used as an effective basis for comparison.

The table in Figure 10.14 shows that the Best Three Warehouses—Juice Inbound solution (Scenario C) yielded a total cost of $27.2 million. Even though we know that the models are not optimizing the same network objectives, we also see that the outbound cost associated with this solution (Scenario C) is significantly higher than the outbound cost for the solution without the consideration of inbound transport costs (Scenario A). We expect these results as the model is adjusted to now trade off inbound versus outbound costs, the result of which yields an understandably higher outbound cost.

	Scenario A	Scenario B	Scenario C
	Best 3 Whs- No Inbound	Original 3 Whs- With Juice Inbound	Best 3 Whs- Juice Inbound
Inbound Costs	$0.0	$15.3	$11.8
Outbound Costs	$10.3	$12.5	$15.4
Total Cost	$10.3	$27.8	$27.2
Avg Distance to Stores (miles)	115	154	202
Locations Picked	Minneapolis Chicago Columbus	Minneapolis Chicago Columbus	Milwaukee Toledo Louisville

Figure 10.14   Cost Comparison of Scenarios with and without Juice Inbound

The scenario results depicted in Figure 10.14 also show us that Scenario C results in a total cost only $0.6 million lower than the comparable original scenario utilizing the Minneapolis, Chicago, and Columbus warehouses but now also including inbound transportation costs (Scenario B). Inbound costs are lower in Scenario C than in

Scenario B, revealing that Scenario C focused more on reducing inbound costs in relation to outbound costs in order to reduce the overall network transport costs. This is where the importance of analyzing both inbound and outbound costs simultaneously comes into play.

We continue our analysis of the impact inbound transportation on our solution, but will leave that to the interested reader to explore in more depth.

The Value Grocer's case study has shown us the important of multiple product groups for more accurate costs (products fill up trucks differently), for different service requirements (frozen products), and for different sources (apple and orange juice).

Modeling Bills-of-Material (BOMs)

When a network design model includes raw material suppliers, the location of plants, and the decision of what product to make where, you will often need to include a bill-of-material (BOM). The BOM tells the model which raw material products are needed to make which finished good. That is, the model does not simply send the same product from the original source to the customer. The product changes form within the model as it moves from a raw material source to a plant and then on to warehouses or customers.

From a network modeling perspective, the BOMs are usually modeled as a combination of ingredients and processes depending on the scope and complexity of the manufacturing analysis. That is, you do not need to include every item listed in the actual BOM and need to think about how to model this effectively.

From an ingredients perspective, it is important to model only key ingredients that will impact the decisions on production sourcing based on costs and capacities. The BOM associated with more complex products may be quite lengthy and include many minor components (like nuts and bolts) that have no impact on results of the network design model. Therefore, our decision on what components or ingredients to include in a model should be based on the following:

- **Contribution of Ingredient to Overall Product Cost**

    The focus should be on components that make up a large portion of the overall material or production cost of a product, and exclude components with low unit cost impact. For example, screws, nuts, and bolts may not be modeled, as they represent a very small portion of the product cost, and the change in production location will likely not have a notiticable impact on inbound costs for these components. Packaging materials may often fall into this category as well.

- **Ingredient Sourcing Constraints and Its Impact on Finished-Goods Manufacturing**

  If there are specific components with sourcing constraints from a transportation or capacity perspective, such components should be included in your modeling. For example, key components that may be sourced only from a West Coast supplier may potentially impact the decision on producing the dependant finished-goods product in a West Coast plant versus East Coast plant. Similarly, any constraints on supply of ingredients such as availability based on seasonality or capacity constraints from specific vendors should also be considered. For a sugar manufacturer, for example, the sugar beets are harvested only during certain months of the year and must also be converted to raw sugar within a short time frame after harvest before they become unusable. Therefore, incorporating these timing constraints and relative importance of proximity to the source of the sugar beets is essential to any optimal solution in their network models.

Similarly, in terms of the production process, your focus should be on the modeling of only the key processes based on the following factors:

- **Impact on Overall Throughput Capacity (i.e., Bottleneck Process)**

  Any given manufacturing process is typically made up of a series of production steps, each one performed on a specific type of production equipment. From a modeling perspective, the focus should be on the key processes that impact the overall manufacturing throughput for a given set of products through the plant—that is, the focus should be on the main bottleneck processes only, because this will directly impact how much product can be sourced from a plant overall. If we take the case of the manufacturing process for chewing gum, the bottleneck is usually within the sheeting process, in which the mixed gum (in paste-type form) is extruded and pressed through sheeting equipment to form thin sheets. This process is typically considered the bottleneck process, so the total volume of finished goods chewing that may be produced is largely dependent on how much sheeting capacity this equipment provides at each plant. As a result, this process should be modeled in the network model as part of the BOM.

- **Impact on Key Capital Decisions**

  It is important to include processes that require expensive capital equipment and drive the decisions on what products to make in each plant. Note that this is focused on equipment that is directly tied to the production process for specific products, rather than equipment that is common to all products. For example, there may be certain tanks used to store byproduct liquids for disposal which are required for a production process but are not directly tied to any specific products. In other words, they are required for all products, and therefore will not influence which products are made at this location.

Bills-of-Material Example—Beer Manufacturing Process Modeling

To better understand the concept of bills-of-material, let's look at the example of beer manufacturing. In this example, a beer manufacturer operates a network of four breweries across the country and is looking to optimize the production of finished-goods products (bottles, cans, kegs in various sizes and packaging types) across the four breweries. We will analyze the ingredients that make up the finished goods and the production process for the manufacturing of these beverages and packaging types, and convert this into a suitable schematic for modeling.

Many different ingredients are required in the production of beer. However, in general the key ingredients include water, barley, hops, and yeast (as shown in Figure 10.15). These are the key components from a production and product cost perspective but we need to carefully evaluate which ingredients are appropriate for modeling.

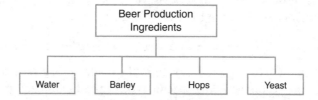

Figure 10.15   Overview of Key Production Ingredients of Beer

Let's start with water. This is the most prevalent ingredient in beer in terms of weight and proportion. However, water is typically procured local to the brewery and usually doesn't require shipping into the brewery from distant locations. So although this is a key ingredient, it does not add value as a component in a network model in this case because the water is always available to be sourced locally and at a relatively low cost to the overall process.

Barley and hops, however, are both typically shipped into the brewery from farms in ideal growing locations. Some BOMs may also utilize specific types of barley or hops in each specific type of beer. In this case we must model each of these ingredients in order to properly include the impact that their sourcing will have on inbound costs when determining where each finished good should be produced.

Lastly, yeast is probably the most critical ingredient in this process. The inclusion of a relatively small amount of yeast enables the production of alcohol from the sugar in the barley. Given the size and volume of yeast used in production, though, it is an ingredient that can be easily moved and shipped throughout any network at relatively low transportation cost. As a result, yeast is another ingredient that can be excluded from the network model, because it will not have a significant impact on our finished-goods production decisions.

Although we typically think about just products when modeling a BOM, we also need to consider the different steps required to make an item. Figure 10.16 shows the basic manufacturing steps for beer.

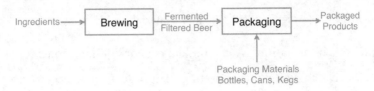

Figure 10.16   High-Level Overview of Beer Production Process

From a manufacturing perspective, this process can be broken into two main categories: brewing and packaging. Note that the brewing process is actually made up of several steps, including malting, milling, lautering, boiling, fermenting, conditioning, and filtering. However, from a modeling perspective, we want to focus on only the key processes that impact the overall throughput of beer production at each brewery. Each of the brewing process steps mentioned previously has its own production rate and throughput depending on the type of beer being produced, but they can collectively be grouped and modeled as an overall process in a network model.

After the beer is fermented, conditioned, and filtered, it is available for packaging into various types of packages—that is, bottles, cans, and kegs. The packaging process also requires additional product inputs in the form of packaging materials—that is, various sizes of empty glass bottles, aluminum cans, and kegs.

These packaging materials are sourced from specific vendors and shipped from each of their associated plant locations. More specifically, 12-ounce bottles may be sourced from one bottle manufacturing plant, while 16-ounce and 24-ounce bottles may come from a different plant. This inbound material sourcing will definitely impact the optimal production location for each of our finished-goods products, depending on where the packaging capability exists within the network.

In Figure 10.17, we show a sample schematic for modeling a BOM for several finished-good beer products in a network model. This BOM includes a combination of ingredients (hops, barley, and packaging materials) as well as production processes (brewing and packaging). This example could be expanded even further to include variations in different types of hops or barley that correspond to different types of beer as well.

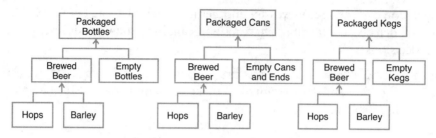

Figure 10.17   Bills-of-Material for Model

When we have this information, we can build a network design model to incorporate it.

Whether it's a network model incorporating the production of beer or one optimizing the production of computer chips, we can clearly see that the inclusion of production modeling requires us to not only differentiate our finished-goods products but also include raw material, component products, and the steps in the manufacturing processs within our analysis.

Lessons Learned from Adding Products

In this chapter, we learned the importance of modeling multiple products as well as the key drivers that make this a necessity for producing accurate results. These drivers include the following:

- Variations in logistics and storage characteristics that require us to be more detailed in our product definition in order to accurately capture the appropriate transportation and storage measures across the model.

- Products with differing service-level requirements require different modeled products in order to assign the appropriate transit time constraints as well as associate the appropriate transport mode and cost to each customer point.

- Variations in production sources and capacity constraints on the amount of product each source may provide require specific production definitions in order to track where each product may be sourced from and how much in an optimal network structure.

- Important raw materials or production steps. When modeling manufacturing operations, the source of the raw materials and steps in the manufacturing process may influence the results.

When additional product families are modeled, the solution can be driven by different products. For example, some product families may drive one set of facilities, and another set of products a different set of facilities. For example, if it is important to be close to the source of products, then the model may be pulled to locate facilities close to the source of Product A and also pulled to locate facilities close to the source of Product B. Adding products creates another internal trade-off that the optimization engine must balance.

The concept of modeling specific products within a network model is important for any company to understand. Not understanding this concept could leave a modeler with inaccurate results due to an inability to define all necessary costs and constraints specific to product groupings. Conversely, modelers may assume that product definition is *always* required when in actuality a lot of time could be saved while producing the same results for a model that doesn't include any of our previously mentioned key drivers.

End-of-Chapter Questions

1. A popular online shoe retailer has been successfully servicing their customers across the United States with the most popular brands of shoes from tennis shoes to high heels for the past five years. The retailer sources these shoes from a wholesale shoe distributor and therefore receives all their shoe orders from the closest distributor stocking location. Due to the bulky packaging required to ensure no damage to the shoes in transit, the transportation cost is always volume dependent (or can be tied back to a per-carton cost).

   The supply chain modeling team has previously modeled a generic volume of shoes flowing through their network to determine optimal warehouse locations. This year the company has decided to expand their retail footprint by beginning to offer the sale of designer formalwear such as suits and dresses as well. These products arrive in the supply chain on hangers and can't be folded and shipped in cartons. The modeling team has determined that these products must be modeled as separate products within the network design.

   Explain why this is a good decision in terms of transportation and warehousing cost differences.

2. Describe how you would model the associated products in each situation:

   a. A German chemical company produces five different liquid chemical products (Liquid Products A through E for this example). These liquid products are often sold together in fifty-litre containers to regional manufacturing plants that in turn produce different blends of cleaning fluids then sold in the consumer market. Liquid Product B, however, is also a special chemical, essential to assisting with the emergency cleanup of oil spills around the world.

   The modelers have decided to model two product groups in this case, one called regular and one called emergency. The regular product grouping consists of Products A through E, including the orders for Product B not for emergency purposes. The emergency product grouping consists of Product B, orders related to the emergency purposes only. Why is this decision appropriate?

b. A Canadian tire distributor sells thousands of different automobile tires to privately owned car-repair garages across the country. The distributor locations receive truckload deliveries of tires weekly from a single wholesale company operating one central distribution center in Canada. Each tire has a specific tread and thickness and thus requires a specific part number during ordering. The distributor has recently decided to expand its footprint to servicing markets in the U.S. and therefore is looking to complete a network design study to determine the best locations for two additional distribution centers within the northern U.S.

Why is this team able to model their network design problem with just one product group?

If this distributor decided to switch to sourcing its tires directly from the tire manufacturers instead of the wholesaler, they would then need to switch to include numerous product groupings in their model. Why is this? What would be the basis for these new product groupings?

3. Weighing Out Versus Cubing Out—Mini Case Study

Value Grocers has decided to expand its model to include product lines such as packaged deli meats and cheeses, crackers, potato chips, canned nuts, yogurt, and milk. You can find the detailed storage and logistics characteristics for each of our new product groups along with a sample model and detailed instructions for further analysis within the Value Grocers Expanded Product Line Exercise.zip file on the book Web site.

a. In the first model built, found in the file referenced above, we use only the weight of the products to determine transportation costs. What are the resultant transportation costs in the results of this model?

b. Let's do further analysis to see how accurate these transportation costs might be. Using the Value Grocers Expanded Product Line Measurements.xls to start, complete your own calculations to determine which of these product groups will "weigh out" versus "cube out" (hit the max capacity based on weight or volume limits first) in real-life transportation costing. Remember that a single truckload of product can transport up to 40,000 pounds or 2,000 cartons of product.

i.　Which products will weigh out?

　　ii.　Which products will cube out?

　　ii.　What is the resultant per-carton transportation cost for each product grouping based on this? What would it be if we didn't differentiate between the products?

c.　Back in our model, let's adjust the model to include both the weight and the volume characteristics of each of our products, as well as apply both capacity constraints on the truckload carrier costs.

　　i.　How does this impact our overall transportation costs within the model?

　　ii.　Does this cause the model to select different optimal warehouse locations?

　　iii.　Do certain product groupings tend to come from certain warehouse locations, or are product groupings fairly evenly distributed across the selected warehouses?

4.　Product Sourcing—Mini Case Study

An Australian gold mine in Kalgoorlie (on the west coast of Australia) is used to supply gold to ten major jewelry-manufacturing customers on the east coast of Australia. The mined raw gold must be refined prior to delivery to these customers, however. This leaves the mining company with a need to determine the optimal location for this refinery. The mining company currently has two potential locations it is considering. The question is, should the refinery be located closer to the mine or closer to the customer base? To determine this, we must model the gold as both a raw material (raw gold) and a finished good (refined gold). A bill-of-material must also be modeled in order to determine the amount of raw gold required to produce one kilo of the resultant refined gold product used to service customer demand. The base model, as well as further detailed instructions for completing this study, can be found within `Australia Gold Mining Product Sourcing Study.zip` file on the book Web site.

a.　In scenario 1 we define the BOM to mimic the fact that only 1.1 kilos of raw gold are required to produce 1 kilo of refined gold that can be used to satisfy customer demand. After reviewing the model setup, run the scenario and review results.

　　i.　Which refinery is selected?

　　ii.　What is the average distance traveled from mine to refinery? Refinery to customer?

b. Let's create a second scenario but alter the BOM. Now the raw gold has a lesser quality and it requires 2 kilos of raw gold to produce 1 kilo of refined gold.

   i. Does this change which refinery location is optimal?

   ii. What is the new average distance traveled mine to refinery? Refinery to customer? Explain how the differing results in these two solutions make sense in terms of the amount of product moving on each type of lane (mine to refinery, refinery to customer) and the associated transportation cost changes for each scenario.

5. Bill-of-Material Modeling—Mini Case Study

   Let's consider the case of a chewing gum manufacturer that is interested in optimizing its manufacturing network and sourcing strategy. The products include sugar-based and sugar-free sticks of gum. Following is a quick description of the ingredients and the manufacturing process:

   - Key ingredients include these:

     - Gum base—Sourced from one of two plants worldwide

     - Sweetener—Sugar or sugar-free substitute, sourced locally from regional suppliers at local prices

     - Flavor additives—Very low weight, sourced from one flavoring plant in Seattle, Washington, for all gum manufacturing plants

   - The manufacturing process starts with the use of mixing equipment to combine all ingredients into either sugar or sugar-free paste.

   - The mixed product is then converted into thin sheets using an expensive piece of equipment called a sheeting machine. The sheeting machine also requires special parts to handle sugar versus sugar-free paste.

   - In some plants, there are sheeting machines dedicated exclusively to each type of product (sugar, sugar-free).

   - The sheeted products are then dried and cut into sticks.

   - The dried sticks are then sent for final processing by packaging machines into any of several package types.

   Let's discuss the modeling of this BOM in terms of both the ingredients and the production processes.

   a. The modeling team has decided to exclude the gum base ingredient from the BOM in their model. Why might this cause inaccurate results in their solution?

b. Based on what you know about the ingredients and production processes described previously, does each instance of the product (semi-finished or finished) need to be distinguished by their sugar or sugar-free nature? Discuss the specific effects on both modeling the production process and demand.

c. Is it essential to include the "mixing" step in the production process within the model? What might be the effect if it is excluded?

d. Is it essential to include the flavor additives in the model? What might be the effect if it is included or excluded?

11

MULTI-OBJECTIVE OPTIMIZATION

So far, we have discussed trying to improve just one component of the supply chain. That is to say, we have talked about minimizing the total distance that our goods have to travel, or the total supply chain cost over some length of time. We have framed strategic network design this way for three reasons:

1. Historically, strategic network design tools have always presented the user with a single goal. The majority of tools (although certainly not all of them) currently give you the power to minimize (or maximize) just one objective function at a time.

2. The mathematical tools that underpin strategic network design are oriented toward optimizing a single objective.

3. Optimization of a single goal is difficult enough to understand without the added confusion of considering multiple simultaneous objectives.

That said, the reality of strategic network design problems is that an analyst naturally gravitates toward more than one objective. For example, it is quite natural for a network design study to recommend the expenditure of a certain amount of upfront infrastructure costs in order to save money over a long time span. The analyst might conclude that significant savings can be realized by offshoring production to the Far East, and that two large West Coast seaports will be a key element of this plan. Although these warehouses will be expensive to build, they will more than pay for themselves over a ten-year time span.

Or will they? As this book is going to press, the price of electricity and natural gas (i.e., industrial energy) is much cheaper in the United States than it is almost anywhere else in the world, and Chinese wage inflation is eroding the cost savings of offshored labor. This combination is contributing to a trend of "reindustrialization" of the American Rust Belt. If this trend continues, a supply chain built just recently with an eye toward offshore production might completely lose its competitive advantage, and require reorienting toward more domestic production.

Of course, these sorts of trends are difficult to predict (if we could predict the price of energy, we could skip supply chain planning and make a fortune on the futures market). However, it is quite reasonable to consider limiting the amount of "upfront" fixed costs that our plans recommend. That way, if conditions change dramatically, our supply chain will not be locked into long-term lease payments on facilities that no longer make sense.

This naturally leads to a conundrum. We want to save money over a long planning horizon, because the most successful organizations invest in long-term success. However, we also want to reduce our fixed-cost capital commitments, because such obligations limit our ability to react to significant changes in the global economy. We have two reasonable goals that appear to be in irreconcilable conflict.

This sort of intellectual crisis isn't limited to "upfront costs versus long-term savings." It can also occur when one considers the trade-off between facility costs and customer-service level. When a customer is a long distance from the warehouse or plant that supplies it, the customer-service level will inevitably suffer. Sooner or later, the customer will experience a surge in demand that exhausts its in-store stock. If the supplier is nearby, an emergency restocking request can easily be accommodated. If the supplier is a long distance away, such a request might be impossible, or exorbitantly expensive, to fulfill. Thus, we might want to insist that every customer be sourced from a nearby facility.

However, such a requirement could radically increase the cost of our supply chain. In the most extreme case, we might end up building ten plants and 70 warehouses in order to service 120 customers. Such a large facility building and leasing expense could be quite significant, and completely dwarf the benefit from superior customer service.

Intuitively, this insight has a certain appeal. Customer service isn't free. If we want 50% of our customer demand to be within 100 miles of a supplier, it might very well cost us more than a solution that places 40% of customer demand within 100 miles of a supplier. Although we can't definitely answer which of these two solutions is "best," it's worthwhile to know the exact nature of the trade-off between these two choices. Even better would be to generate an array of worthwhile choices, each one representing a different trade-off between customer-service level and total cost.

Luckily for us, mathematical optimization has developed the tools to address the situation. The key is to recognize what it means for a solution to be worthwhile. More precisely, we will describe what would disqualify a solution from consideration, and then the set of worthwhile solutions will be those that cannot be disqualified.

Consider the "upfront cost versus total cost" conundrum described previously. Suppose an analyst generates three possible solutions to choose from (see Figure 11.1). The first solution costs $10 million upfront and $100 million overall; the second, $12 million upfront and $90 million overall; and the third, $11 million upfront and $100 million overall. The first solution is compelling because the upfront cost is so low, and the second is appealing because it represents a significant long-term savings. However, the third solution has no specialty to hang its hat on, as it has the same long-term cost as the first, but with a higher upfront penalty. Thus, we can safely disqualify the third solution from consideration, while the first two appear to be worthwhile.

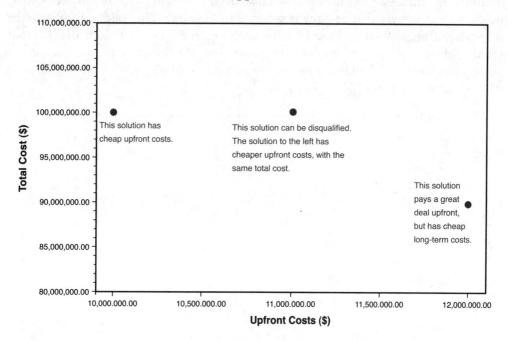

Figure 11.1   Chart Showing Trade-off between Total Cost and Upfront Costs with Three Solution Points

More formally, mathematicians describe a multi-objective solution as Pareto optimal, if it is impossible to generate a "disqualifying" solution that improves on one objective without penalizing the other. In our example, if the first solution is Pareto optimal, there must not exist a solution with upfront cost less than or equal to $10 million and total cost strictly smaller than $100 million, and similarly there must not exist a solution with upfront cost strictly less than $10 million and total cost less than or equal to $100 million. Were such a solution to exist, it would be clearly superior to our first solution, as it would achieve at least as much long-term savings with a cheaper upfront cost, or it would generate a larger long-term savings with no larger initial outlay. A solution is thus Pareto optimal if no other clearly superior solution exists.

However, there might be many, many Pareto optimal solutions. For example, there might be four Pareto solutions, each with the following (upfront cost, total cost) values: ($10.5 million, $100 million), ($12 million, $90 million), ($15 million, $87 million), and ($19 million, $85 million).

These four solutions (see Figure 11.2) each represent different trade-offs between upfront cost and long-term savings, and thus, from a purely mathematical perspective, we can't definitely say that any one is superior to the other. That said, it would not be surprising if the flesh-and-blood humans analyzing these four solutions decided that the second one is the pick of the litter. Its upfront cost of $12 million is quite close to the best upfront cost of $10 million, and its long-term cost of $90 million is quite close to the best long-term cost of $85 million.

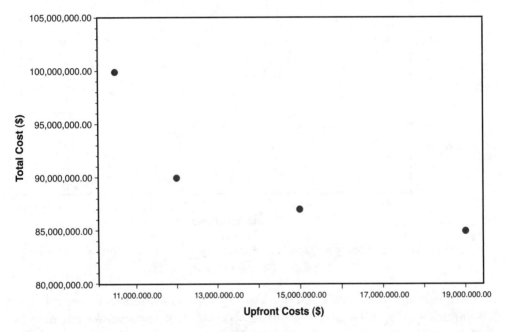

Figure 11.2   Chart Showing Where the Four Pareto Optimal Solutions Sit on the Graph

Historically, strategic network design modelers have used "what-if" analysis to discover solutions that, though not provably Pareto optimal, are at least "close" to Pareto optimal solutions. That is to say, with the example previously given, a user might try to minimize total cost by optimizing three scenarios with three different upfront cost restrictions: "upfront cost <= $10 million," "upfront cost <= $15 million," and "upfront cost <= $20 million." These results might play out as follows.

Scenario 1 is infeasible, because the upfront restriction was too tight. That is, there is no solution whose upfront cost is less than or equal to $10 million. Ideally, this infeasibility would be well diagnosed by a sensible error message, but in general, infeasibility diagnosis is quite hard.

Scenario 2 might generate our third Pareto solution with upfront = $15 million and total = $87 million.

Scenario 3 might generate a solution that is "close to" Pareto; that is, upfront cost = $20 million and total cost = $85 million.

Figure 11.3 shows how these three scenarios play out.

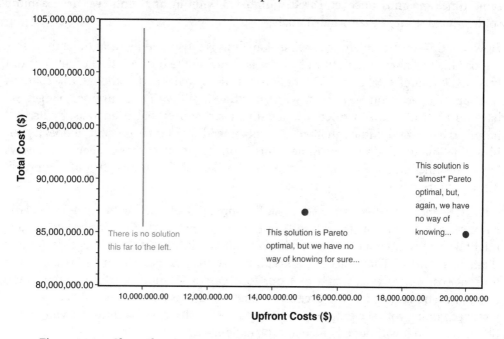

Figure 11.3 **Chart Showing a Typical Result with "What-If" Analysis, Instead of Real Multi-Objective Analysis**

We can thus easily see the imperfections of this time-honored what-if analysis strategy. The first problem is, simply, how do you determine the restrictions to apply to each objective? Clearly, our first scenario didn't add much value, because its upfront cost restriction was unrealistically ambitious. Had we restricted upfront cost to $10.5 million, we would have discovered our first Pareto solution. However, when we're starting from scratch, it might be no easier to realize that $10.5 million is a good what-if restriction than it is to determine next week's lotto numbers.

The second problem with what-if analysis is that the solutions that it generates need not be Pareto. Consider the third scenario. Our solver engine minimized the total cost in this case while obeying the restriction that the upfront cost not exceed $20 million. In doing so, it discovered a solution that has total cost of $85 million and an upfront cost of $20 million. However, a clearly superior solution exists with the same total cost, and a smaller upfront penalty (Pareto solution number 3, which has total cost of $85 million and an upfront cost of $19 million). Why didn't the solver generate this better solution instead? Because it was under no obligation to do so. Like any good computer program, it did what was asked of it—found the cheapest solution that doesn't overrun $20 million in initial outlays. If we want our optimization tool to consider both goals at the same time, and thus uncover Pareto number 3 without any hints, we need a more sophisticated strategy than what-if analysis.

However, this third problem with what-if analysis is illustrated by what these three scenarios failed to discover. Specifically, what is arguably the best solution of all (upfront cost = $12 million, total cost = $90 million) was never generated. Although this solution will appear if we rerun Scenario 1 with an extra $2 million in the upfront budget, we have no particular reason to choose just this restriction. What-if analysis, by its very nature, is haphazard. Although it will give us some idea of what's possible, it does a poor idea of telling us what's not possible. Thus, a practitioner of what-if analysis is always left with some lingering doubt that the "Goldilocks"-perfect trade-off might appear if just one more scenario were optimized.

Luckily, better automated strategies exist. For some models, we might ask a particularly advanced tool to generate the full suite of Pareto optimal solutions; because such a suite might be large, our clever tool would likely graph the Pareto results, rather than present them as a flat list. That is to say, a graph might chart total cost on the y axis, chart upfront cost on the x axis, and plot our four Pareto solutions as dots on this graph (along with other dots for whatever other Pareto solutions exist). This is in fact what we have been doing with the graphs in this chapter—the strategy is so intuitive you probably understood it without any formal explanation.

However, this is, in general, an overly ambitious strategy. For some models, the number of Pareto solutions is so vast, with the differences between them so minor, that generating all of them might be pointlessly time-consuming. For other models, solving just one objective might be so time-consuming that it might be more realistic to simply remove large swaths of the upfront versus total cost graph from consideration. That is to say, it might be more worthwhile to graph what is and isn't possible, and then let the user decide which sections need deeper consideration.

This is the strategy that commercial, off-the-shelf tools are following. A user will specify two objectives (i.e., total cost and upfront cost), and the tool will automatically draw two lines (or frontiers) on a bi-objective graph. This graph is seen in Figure 11.4. The upper line shows "what's possible." Although it might not be possible to create a solution that is "under and to the left" of this line, it's certainly true that any solution that plots "above and to the right" can be improved upon. Similarly, the lower line shows "what's impossible." Although it might be possible to generate a solution that is inferior to a point on the lower line, it is impossible to generate any solution that plots "under and to the left" of the line as a whole. Thus, the "achievable frontier" (the upper line) and the "impossible frontier" (the lower line) can build intuition and help focus what-if analysis even in the early stages of optimization, when they are quite far apart.

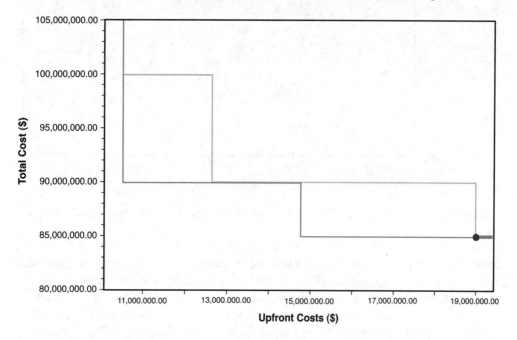

Figure 11.4   Chart Showing Achievable Frontier (Upper Line) and
Impossible Frontier (Lower Line)

As a multi-objective solve process progresses, it discovers more refined information about both what's achievable and what's impossible. This superior information is conveyed visually by redrawing the two frontiers. As new information comes in, the "achievable frontier" moves down and to the left, representing the discovery of solutions that have cheaper upfront costs or larger long-term savings. As new bounds are deduced, the "impossible frontier" moves up and to the right, representing the discovery that certain (upfront cost, total cost) pairs are unattainable.

For certain models, a user might realistically wait until the two lines completely converge (shown in Figure 11.5). Such a happy result would represent the discovery of the complete set of Pareto optimal solutions. These solutions would lie at the "L" corners, where the bottom of a vertical line segment meets the left end of a horizontal line segment.

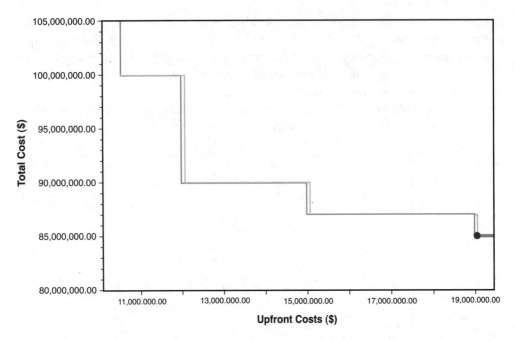

Figure 11.5   Chart Showing That the Two Frontiers Have Almost Completely Converged

For most models, a true enumeration of the full set of Pareto solutions is unrealistic. However, such a perfect enumeration is hardly necessary. It's quite easy to see that when the "daylight" between the two lines has been reduced to a mere glimmer, we have enough information to begin a comprehensive and orderly "what-if analysis" study. To return to our previous example, we don't need the lower and upper lines to perfectly coincide to recognize that a solution very close to ($12 million, $90 million) is possible, and thus to launch a single optimization run with the upfront costs restricted to less than $12.05 million.

Lessons Learned with Multi-Objective Optimization

As we have previously seen, different objectives drive us to different solutions. In many cases, each of these objectives is important.

Multi-objective optimization is a technique that allows you to analyze two objectives at the same time to determine the appropriate trade-off.

End-of-Chapter Questions

1. If you run a multi-objective optimization and generate a Pareto optimal set of solutions, what work do you still have to do to determine the best solution for your supply chain?

2. If two large companies merge, why might you want to run a multi-objective optimization to trade off the total cost of company one versus the total cost to company two? That is, why might this be better than running with just the objective of minimizing the total cost of the combined supply chain?

3. Open the model and file `Germany Store Delivery.zip` on the book Web site. This file contains a multi-objective optimization for which the goal is to minimize the number of depots needed to serve stores in Germany such that as many customers are within 75km as possible. How many depots are needed to reach 50%, 60%, 70%, 80%, 90%, and 100% within 75km? If the management team thought they wanted 100% within 75km, how would this chart help you persuade them that this was not a good idea?

4. Let us revisit the case study from Chapter 10, "Adding Multiple Products and Multisite Production Sourcing," for Value Grocers—the grocery retailer that operated a network of stores in the Midwest region. If you were to run a multi-objective optimization with this case study, what are the various objectives you would consider in this analysis?

12

THE ART OF MODELING

As you've seen throughout this book, network design is about making a decision about your supply chain. When you are in the midst of analyzing data, are talking to different people in the organization, and are worried about the deadline, the most common trap you can fall into is thinking that the goal of the project is to exactly replicate the current situation. This thinking leads to too much effort in collecting data, too much effort in building a perfect model, and, in the end, failing to create a model that can be used to make a good business decision. In other words, if you try to exactly replicate reality, your model will be just as complicated and just as expensive as reality.

Instead, you need to keep in mind that the objective of the model is to make a good decision for the supply chain. This is why modeling is more of an art than a science. You've seen the science of solving these problems throughout this book. The art of these problems is setting up the models so that they can give the business the information it needs to make a decision and take action.

It should be noted that the definition of a good decision almost certainly includes some timeline. Businesses move fast. A great decision delivered too late is much worse than a good decision delivered in time to act. Keep in mind the adage attributed to Voltaire: "The perfect is the enemy of the good."

Of course, generic adages don't help directly with network design. In this chapter, we will explore the following specific areas that will help you build better models:

1. Start by understanding the supply chain.

2. Separate the trivial from the important.

3. Start with small models and iterate.

4. Run a lot of scenarios—don't be afraid to experiment.

5. Don't be afraid of including things in the model that don't exist in the actual supply chain.

6. Models are not a substitute for due diligence and decision making.

7. Optimization will do anything to save a penny.

The next two chapters cover the important topics of data aggregation and running a project.

Understanding the Supply Chain

Before starting your network design project and data collection, it is important to complete a scoping session to ensure that all key stakeholders agree on the exact questions being answered, availability of data, aggregation schemes, and output expectations. This discussion and the decisions that come from it will shape the rest of the project from start to finish.

This session should include representatives from Production, Warehousing, Transportation, Finance, IT, and any associated Project Analysts. Ensuring participation from all associated areas from the start ensures significantly less pushback on solution implementation after you are well into the project.

A good early step is to draw a diagram of the physical structure of the supply chain on a white board to understand the current flow. Drawing this diagram helps the team to better understand the full scope of the supply chain. Different members of the team may have thought about only their portion of the supply chain. Seeing the full supply chain in one diagram is a great way to understand it and a great way to get various people in the organization to agree that it is correct.

For example, a retailer may draw the diagram shown in Figure 12.1 on the white board to show the flow of their product from the vendors to their vendor's warehouses (called DCs—distribution centers) and then their own DCs and on to the stores.

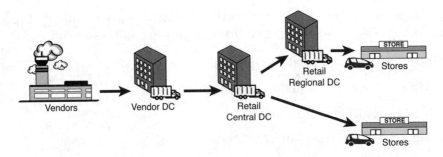

Figure 12. 1   Schematic of a Supply Chain for a Typical Retailer

Note that we have not drawn every store, but the general groups and categories of stores. In this supply chain, the retailer is looking back to its vendor locations.

With this diagram in place, the retailer can start to ask questions about whether they really want to include the vendors and vendor locations. They can ask whether they have missed any important flows in the supply chain. For example, does the vendor ship product directly to the stores, and is that important? If these retailers sell online, someone on the team will quickly realize that this diagram has omitted the online customers. Also, as the team draws the diagram, you might realize that there are different types of stores that require different rules. For example, smaller stores in the city center may receive deliveries from an intermediate depot. Or, if the retailer is a grocer, you may need to include different warehouses for frozen, refrigerated, or ambient warehouse types.

A consumer packaged goods (CPG) manufacturer (or any industrial manufacturing firm) would have a different diagram. For example, a typical CPG company's supply chain may look like the schematic shown in Figure 12.2.

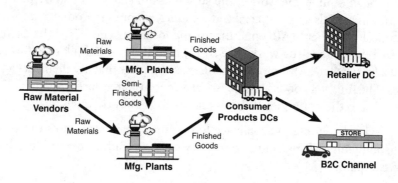

Figure 12.2   Schematic of Supply Chains for a Consumer Products Company

In this case, there is more focus on modeling the manufacturing process. The location of the raw material vendors can play a role in what product is made where and where plants are located. There may be multiple manufacturing steps that are performed at different locations. Also, note that the end customers include both the retailer's warehouses (referred to as the Retailer DC—distribution center) and the B2C (Business to Consumer—or online) Channel. Many manufacturing firms are starting to sell their product directly online. This creates the need for a different type of customer. For example, they may ship in full truckloads to the retailer's DCs, but may ship small-parcel for the online business. Also, in this case, all product goes to a Consumer Products distribution center (DC) before moving to the customers. The team should also ask themselves whether the plants can ship directly to the customers and bypass an additional distribution center.

An industrial manufacturing company, like a chemical company, will have the same general structure except their customers may be other manufacturing companies.

A pharmaceutical supply chain may have yet another structure and it may have plants located around the world. The diagram shown in Figure 12.3 depicts a typical pharmaceutical supply chain.

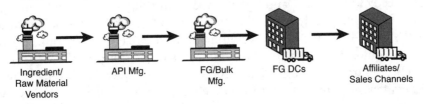

Figure 12.3   Schematic of Supply Chain for a Pharmaceutical Company

A typical pharmaceutical supply chain consists of a series of steps performed at different plants located around the world. The supply chain starts with sourcing of ingredients and raw materials from specialized vendors, shipping products into an Active Pharmaceutical Ingredient (API) manufacturing location. The API is the chemical formulation that makes the drug work. The API is then shipped in bulk to a finished goods (FG) plant that turns the API into pills, tablets, and liquids that are then packaged in the right doses. The finished goods are stocked in a central warehouse (called FG DC in this case) that ships to their regional affiliates and various sales channels.

With the previous examples, you can see how supply chain structures can be very different depending on different industries. They can also capture the unique characteristics of the company.

As you are creating the diagram, you want to ask as many probing questions as possible to make sure you understand the full modeling requirements. We have provided a set of questions, in the text that follows, as a good guide to help get you started. A typical approach that has worked well in the past is to get the group talking by asking everyone to describe their expectations for the project in general. Often many of the questions will be answered during this initial discussion. The provided questions can then be used to help further the discussion and to finalize official scope documentation. Though not all questions need be answered, the more information gathered the better. The output of this discussion also directly guides your next step, which is to begin the data collection process.

The following list of questions will help you define the objective, constraints, and decision variables. These questions are just a list to get you started. You should make sure you ask plenty of follow-up questions and make sure you capture many details. Later, you will need to sort through the answers and separate the important from the trivial (more on this later in the chapter). But, for now, getting detail is important.

1.   What is the objective of your network optimization study? For example, are you trying to:

   a. Reduce transportation costs? Reduce total costs? Reduce costs after a merger?

   b. Reduce the delivery time to customers?

   c. Better use your capacity or add new capacity to cope with growth or a new channel?

2. What are the key decisions you want to make? What specific output should be available after the completion of this project?

3. Which portions of your network may change and which are fixed for the purposes of this study?

4. How much of your supply chain do you need to include in the study? Do you need to include both manufacturing and warehousing in the study?

5. What is the desired project start date? (Completion date?) What is driving the completion date? If the timeline is short, you may not be able to create a detailed model.

As part of the study, you also need to understand the data that must be collected. Often, understanding the data will help you ask better questions from the list just presented. That is, these questions should not necessarily be sequential, but should be iterative. To understand the data, it is often a good practice to focus your questions on the physical supply chain and the different types of products, customers, warehouses, plants, and transportation. Here is a starter set of questions:

1. How many different products or SKUs do you manufacture or distribute?

   a. How do you currently categorize and group these products?

   b. Where do these products come from? What are your options?

   c. Will you need to consider the key raw materials that go into the finished goods? How do you categorize your raw materials?

2. How many warehouses do you have? Where are they located? Where do they get product? Who do they ship to? How are they different from each other (are some for all products, some for frozen products, some for online customers, and so on)? How do you measure warehouse capacity? Do you want to consider new warehouses? What are the fixed and variable costs of the warehouses?

3. How many plants and suppliers do you have? Where are they located? What do they make? Where do they send their products? How do you measure their capacity? Can the same product be made in multiple locations? What are the fixed and variable costs of the plants?

4. How many customers do you have? Where are they located? Is there wide variation in the demand of each customer? How can they receive product? How do you group your customers? Do you have different channels? Do the customers have different service requirements (need to be within 100 miles of a warehouse, need to receive product in three days)?

5. How do you ship product throughout your supply chain? Where do you use rail, truckload, less-than-truckload, parcel? What are your rules for when you use different modes?

Now that you have a working diagram of the supply chain and the details behind that diagram, it is time to start thinking about the model. By now, you know quite a lot (too much, maybe) and it is time to determine what to model.

Separating the Important from the Trivial

Often, the difference between a model that leads to decisions that save the firm millions of dollars and a modeling project that fails is the ability to separate the important from the trivial. If a modeling project tries to get down to every last detail, not leave out a single exception that happens in practice, or include every minute decision, most often it gets bogged down to the point of failure.

When we phrase it like this, it looks easy to avoid this pitfall. However, in practice, it is not so easy. There will be strong forces in play that will push you into more and more detail. For example, there will be a strong feeling that if the model does not represent the supply chain, we cannot trust the results and we cannot get others to believe our analysis. If we simply give into these forces and feelings, and try to model everything, our models will fail. Instead, we need to understand that these forces exist and work to address them and still build a practical model.

The art of modeling requires that our models capture the important aspects of a problem, that they don't get bogged down in the trivial, and that we and others can trust the results.

There is no magic to separating the important from the trivial, and you may not even be able to do this prior to digging into the problem. Although the list is not comprehensive, here are some tips to help you develop better intuition for separating the important from the trivial:

- **Realize that there are different types of decisions and aspects to the problem**—Just realizing that there may be important and trivial decisions is a good first step. This allows you to realize that you can separate out different decisions and then you can decide which ones to include or exclude later.

- **Remember the 80/20 Rule**—This common rule of thumb helps you realize that 80% of your sales come from 20% of your customers, 20% of your products account for 80% of the business, 20% of the supply chain changes will lead to 80% of the savings, and so on. It is not a hard-and-fast rule, but helps remind the team that if you are spending most of your time trying to understand an aspect of the business that accounts for 1% of the supply chain costs, you may not be focused on the right things.

- **Exceptions are captured in an optimal baseline**—Almost every supply chain has many different exceptions and special cases. These can be very difficult, if not impossible, to capture. One thing to keep in mind is that we take care of some of this problem with the optimal baseline. In the optimal baseline, we are coming up with a solution to the supply chain that assumes that there are no exceptions. Then, we compare our optimization runs to the optimal baseline. It is not that we assume that exceptions won't happen; it is that we know they will, and that is why we compare back to the optimal baseline. See Chapter 8, "Baselines and Optimal Baselines," for more details.

- **Address the areas the model does not cover**—Even if we have removed the trivial aspects from the model, it may still be in your best interest to make sure you cover these aspects. Covering it does not mean modeling it. It could mean adding in the impacts after the model runs, or it could mean just discussing this aspect during the presentation of the results.

Start with Small Models and Iterate

We have found that starting with small models and incrementally creating ever-more-complex models is an effective way to get to a final model and final decision.

The outline of this book provided you with different building blocks you can use to start simple and add complexity. You saw how you could start with the simple center of gravity model to determine the best supply chain considering just service. Then, you could add complexity by adding one element at a time like

- Capacity
- Transportation costs
- Facility costs
- Multiple echelons

The benefits of this approach are these:

- It allows you to test your data and model incrementally.

- It enables you to better explain your final results if you know the impact that each new element has on the model.

- It allows you to separate the important from the trivial by being able to see the effect of different elements of the model. If something has no impact, you might consider leaving it out.

- You may find that you come up with your answer before you have to build the full model. That is, as you iterate and build more detailed models, you may find that you have a model that answers your questions with less complexity than you were planning on adding.

Run a Lot of Scenarios—Don't Be Afraid to Experiment

The benefit of creating a model of your supply chain is that you can test different ideas, different data sets, and different strategies. It is important to note the alternatives to building a model before making a decision:

- Just implement the change and see what happens. This has the advantage of being very real—you know what the results are. Of course, this comes with the big downside that if you get it wrong, the firm could be financially wrecked.

- Use intuition and judgment. This has the advantage of being quick, but the big downside is that it does not take advantage of data nor does it allow you to sort through the myriad of possible solutions.

Many firms rely on some version of these alternatives. It should be no surprise that the resulting supply chains leave a lot of money on the table or don't support the overall business strategy.

With a model you have the ability to do much better, but only if you make use of one of the big advantages of a model—the ability to run a lot of scenarios. Often these scenarios are called "what-if" analysis. You can test *what* would happen *if* something were changed. And that "something" is up to your creativity.

After you build a model, it is relatively easy to run different scenarios. Besides the debugging benefits we discuss later in this chapter, running scenarios allows you to gain a deeper understanding of your supply chain and better prepare for future possibilities.

Most supply chains are relatively complex with many different costs and parameters. By running the multiple scenarios, you can start to see how changes to one part of the supply chain impact another part, you can see how the variables interact, and you can get a sense of the important variables. You may also get other, unexpected ideas on how to improve your supply chain. One company we worked with wanted only the lowest cost network. However, when they found a solution that was just $3 million more expensive than the lowest cost solution but doubled the number of customers they could deliver to within one-day, they opted to implement the latter solution.

Also, by running different scenarios, you can prepare better for the future. When we are building a structure for our future supply chain, we have to make our best guesses about demand and costs in the future. So it is always best to test what would happen if demand went up or down, if it varied by region or part of the world, or if we lost or gained a new customer. It is also good to analyze what happens if costs change. For example, what would happen if the price of oil were to drop or rise significantly?

Part of the art of modeling and coming up with good answers is to not only test these varieties, but then come up with better answers. One way to think about a better answer is to come up with a robust solution. A robust solution is one that is reasonably good across a wide range of changes to the data.

For example, we worked with a firm that had designed an optimal supply chain that reduced transportation and warehousing costs by $4 million with no changes to service level. They were ready to implement this solution. However, when we analyzed the solution, we realized that the savings were being driven by one very large customer. If this customer was lost or its demand shrank, the new network would have been $500,000 more expensive than the existing structure. However, by coming up with alternatives, they found a solution that would still save $3.2 million if the new customer stayed and $2.8 million if they lost the new customer. In this case, the $800,000 of savings that they gave up was good insurance and made for a more robust solution.

Don't Be Afraid of Including Things in the Model That Don't Exist in the Actual Supply Chain

When you realize that the underlying mathematical optimization engine solves a problem in a systematically ruthless way, you can take advantage of this fact to help you make better decisions and make more efficient models.

For example,

- One way to trick the optimization into maximizing the number of customers within 400 miles of a warehouse is to set up a transportation rate structure so that all customers within 400 miles cost $1 per mile to service, and those outside of 400 miles cost $10 per mile to service. So if one customer is 399 miles away from the warehouse, each truckload costs $399. If it is 401 miles away, the cost is $4,010. If you don't put any more costs into the model, the optimization engine will work very hard to get as much demand as possible within 400 miles. The costs of $1 and $10 are purely to guide the optimization to make the decision you want it to make.

- In our discussion of transportation costs, we mentioned that some transportation rates could be quite flat. That is, it may cost the same amount whether the load is going 10 miles versus 200 miles or 400 miles versus 800 miles. This can happen when you ship small-parcel, or ship with a third party with an extensive network in place and the marginal costs to them are just the extra touches—the vehicles are already moving. In this case, the customers may not be assigned to the closest warehouses. One quick trick is to add a small surcharge to all such shipments of something like $0.01 per mile. This cost will be trivial in the overall scheme of the model, but it can be enough to break the tie when the rates are the same between two warehouses and assign the customer to the closest facility. No such cost exists in the real world and you need to make sure that it gets subtracted later (unless it gets rounded out when you report on significant digits).

- For some models focused on the U.S. with significant imports from Asia, you may model the supply location as the port of entry. For example, you model the supplier or plant in Long Beach, California. There is no such plant there, but it helps simplify the modeling without impacting the decision.

This list could go on and on. The key idea is that if you understand the underlying optimization engine, you can set up structures to get the model to help make the key trade-offs you need to make.

Models Are Not a Substitute for Due Diligence and Decision Making

All through this book, we have been talking about the value and importance of network modeling, so this point may sound like minimizing the value of modeling. However, it is important to remember and remind others about the concept of modeling. A model by definition is meant to be a representation of a real entity, be it a facility or equipment or a supply chain. No matter how detailed or precise (again going by our previous reference to precision in modeling) a model is, it can never represent reality entirely. All network models will have some degree of assumptions and simplified rules. As a result, it is important to use the outputs of the model along with other factors and analysis to make decisions. Network modeling is designed to provide decision support and is not a substitute for required due diligence and decision making. It is important to remember that the results are based on best available data and estimates at the time and may likely change in the future.

We worked on a network study for a company where the final results recommended closing a warehouse in New Jersey and opening a new warehouse in Memphis. The existing cost with the New Jersey warehouse was $10 million and the modeled baseline cost was $9.5 million; and this was good enough for decision making. When the model ran, the Memphis solution in the model came out to $7 million with the baseline demand. After the recommendations were accepted and the implementation was started, the Finance team revised their logistics budget for the following year and set it to $7 million.

What is wrong with this? The $7 million cost estimate was based on assumptions and estimates such as historical demand and nonnegotiated transportation rates. The model was designed to answer questions on the distribution network strategy—the scenario costs are meant to help quantify how much better the Memphis solution is compared to, say, a Chicago solution with a total cost of $10 million. This model was not meant to serve as a budgeting tool, and is not appropriate to use as a substitute for due diligence and analysis that goes with a budgeting process.

Optimization Will Do Anything to Save a Penny

When you set up a network design model and hit the run button on the optimization, the optimization is going to work as hard as possible to find the absolute best solution. If the optimization can find a solution that is one penny cheaper, it will do so.

A good modeler will keep this in mind.

This can impact how you set up the model. For example, you may want to artificially inflate the cost of opening a new plant to ensure that if the model does open the plant, it is a good decision (see the earlier section on including things in the model that don't exist in reality).

This can impact how you analyze results. Specifically, this points back to the need for multiple scenarios (or what-ifs) and due diligence to test whether the model did, in fact, make a decision to save a penny. We have seen that executives have good intuition with regard to this. In our experience, if we show the executives the best four-warehouse locations, they will immediately ask what the cost is of the best three and the best five. Also, if one of your locations is Birmingham, Alabama, they will ask about the cost if it is in Atlanta. They are, in effect, trying to figure out whether the model is recommending one solution over another to save a penny.

Debugging Models

Earlier in the book, we saw how we can set up the model in ways that prevent a successful run. In the chapter on service levels, we saw how we set up a constraint that said all customers had to be within 800km of a warehouse and that we could have only three warehouses. There was no way for both of these things to be true at the same time. Supply chains are complex and the models you build will reflect that complexity. When you set up these models, you want to make sure they are correct. Sometimes the problems are obvious, but smaller, less obvious problems with your model may also lead you astray.

When a good software developer writes a program, they work hard to get the logic correct from the start, they double- and triple-check the logic, and then they thoroughly test the program with many data sets. There is no shame in finding a bug during the development process. Programs are complicated, with many different interactions. This is why the testing of programs is just as important as writing them. In fact, it is important to note that programs are tested during the writing phase. You do not wait until you have a program to test; you should test it throughout development.

Modeling your supply chain is a lot like writing a software program. Of course, commercial-grade network design software is already debugged. But when you set up your supply chain within these programs, you need to make sure it is doing what you want it to do. To ensure that it is working properly, we suggest that you follow the steps a programmer goes through.

To start, you should make sure you understand the key elements of the optimization model you have set up. These elements are explained in Chapter 1, "The Value of Supply Chain Network Design," but are worth reviewing:

- **Objective**—What are you optimizing and how do you compare one solution with another?

- **Constraints**—What are the business restrictions you want to model?

- **Decisions**—What are you asking the model to do and what choices are you giving it?

- **Data**—What data will you load into the model and is the data what you expect it to be?

Taking a page from good software development, you should thoroughly test your model. Our experience has shown that people are hesitant to test until they have a complete model. We suggest you test throughout the process. In this case, testing a model consists of running a scenario. The scenario need not be complete. By testing and running various scenarios, you can test the logic of your model and the quality of the data. And, as a side benefit, you may also learn a few things about your supply chain that are worth investigating.

What are you looking for when you test? When you run a scenario, you are answering a few basic questions:

1. Did the model come back with a feasible solution? That is, did the optimization problem return a solution or did it come back with a message that the model is infeasible (commercial-grade packages can provide some messages as to what the problem may be).

2. Is the result doing something that makes sense? That is, did it pick locations that make sense, does the flow of product make sense? Does the solution follow the business rules we set up?

3. Do the output values look correct? That is, do the costs make sense?

Even if these three questions can be answered in the affirmative, you still want to test other possibilities. Remember, these models are complicated and there could be hundreds of millions of different possible solutions. You want to make sure that bad logic or bad data is not preventing other better solutions from surfacing. Running a lot of scenarios and forcing different solutions allows you to look for hidden problems that do not immediately surface. If during these tests you come up with a negative answer to one of the preceding questions, you need to dig deeper into your model.

It is also worthwhile, in the early stages of model development, to develop a feel for the nature of optimization. Remember, at the heart of a network design tool is a purely mathematical engine that will inexorably seek the optimal solution. Model analysis and troubleshooting sometimes resembles reverse electronic psychoanalysis, as the flesh-and-blood modeler attempts to deduce the motivations that drive her digital advisor to make such odd decisions. It is better to encounter these conundrums with the simpler and faster models that arise early in the planning process. Thus, the modeler should dedicate some time to playful experimentation throughout the development of industrial-sized models, in order to familiarize herself with the "personality" of the optimization problem she is building.

Fixing Infeasible Models

If you are at the point where you cannot positively answer the first question, you obviously do not get a chance to evaluate the second two questions. So when the model does not run, what do you do?

First, a quick note about infeasible problems: As you have seen throughout this book, these problems are solved with mathematical optimization. If you simply write a mathematical optimization program, and feed the model data and constraints that conflict with each other, then the model will simply return an answer to you which states that the problem is infeasible. It would be up to you to try to determine what went wrong.

Fortunately, commercial-grade network design software can help you pinpoint some problems. First, you know that the software itself has been tested, so you need to just focus on your model and data. Commercial-grade network design software has the ability to pick up some types of problems automatically. For example, if you load a demand of 100 and a total supply of 90, you will get a message that demand is greater than supply. There are a lot of cases like this. However, not everything can be pinpointed without deeper analysis.

You should be aware that not all contradictions can be found in such an easy way. In the example we mentioned at the start of this chapter, every customer was within 800km of at least one warehouse and most customers were within 800km of several warehouses. So you need to run the optimization to sort through the combinations to see whether there is some choice that can meet both restrictions. In more complex models, you may have many such potential contradictions and need the model to sort the combinations.

So for these more complex models, you may still need to do some debugging to get to a feasible model.

One technique we have found to be helpful is to run the model in a mode that allows you to find the least infeasible model by relaxing certain constraints. Some commercial-grade software will have this feature since it requires a difficult reformulation of the problem. This then allows you to determine how close to feasible you are. For example, are you able to meet 99% of the demand or 5%?

Moreover, a commercial-grade tool will include a "bottleneck" report that helps you identify the blocking constraints that are limiting your model's ability to meet demand. Thus, if a partial solution is generated that meets 85% of demand, but whose only fully utilized facility is the Cleveland warehouse, then you are at least oriented toward a worthwhile starting point in the debugging process. This particular infeasibility might result from the modeler's underestimating the capacity of the warehouse in question, or it might result from some more indirect problem, such as a shortage of shipping choices for Midwestern customers. But it would seem reasonable to conclude that the production capacity and shipping lanes associated with the Mexican production facility are likely not to blame.

Another approach is to create some high-cost dummy facilities and dummy transportation costs that are set up in a way that they will surely be feasible, but come at a high cost. If these facilities or transportation lanes are used, it can help point to potential problems.

Usually, infeasible models are the result of constraints that are too tight, constraints that contradict each other, or complex models with missing links. For example, when we define that we want three warehouses and every customer has to be within 800km, we may have constrained the problem so much that there are no possible solutions.

Constraints that contradict each other usually occur when you combine constraints that force a minimum at some point and a maximum somewhere else. For example, you may specify that a plant make at least a minimum of 100 units of a certain item. But because of other capacity restrictions (maybe not at the same location), it may not be possible to make 100 units at that location. The more complex the model, the harder these contradictions are to spot. Our experience has shown that examining the minimum constraints for problems is a good place to start.

Missing links in the supply chain are easy to spot in simple models, and commercial-grade software can pick this up. The problem occurs in complex models with multiple echelons. In these cases there may be links between all the pairs of sites you are interested in. But, based on the way you set up the model, there are certain products that need to get to certain destinations, but cannot do so because of the way you have set up the different flows.

After you have a model that runs and produces answers, you need to address the next two questions (Do the results make sense? and Does the output look correct?).

Fixing Feasible Models

When a model is feasible but is producing results that do not make sense, you need to investigate. This is true even when the results don't make sense just a little bit. We have seen many models that have returned very counterintuitive results that turned out to be true. But to convince people they are true, you need to understand why the model gave the answers it did.

We worked with a company that had experienced rapid growth in Asia and needed a production plant in the region. Demand was growing fast in all parts of Asia, but China was by far the largest market. At the start of the project, the team thought that a plant in China would be the best solution. However, when they ran the network design model, the optimization runs kept suggesting a plant in Vietnam. The company was convinced that there was a mistake with either the model or the data. We had to investigate what was causing these results. It turned out that Vietnam did have higher transportation and duty costs, but because of the low cost of manufacturing and the lower fixed costs, it was indeed the best solution from a cost perspective.

Not all stories turn out like that. Sometimes the strange results just mean that the model is not set up correctly.

This book cannot list all the problems you might encounter. However, keep in mind that the complex model you are debugging is nothing more than a combination of the different building blocks we talked about in this book. If you can understand each of the individual elements, it makes the more complex model less intimidating. We have found that a systematic way to look for problems is to focus on the main elements of an optimization problem. Each of these elements tends to lead to different types of problems. Let's take each one in turn:

Fixing the Objective of the Optimization

Some problems with models are caused because the objective of the model is not set up like you want or you have not captured some important business aspect of the problem. Remember, the objective function is what guides the mathematical optimization engine and allows the engine to compare one solution to another. The optimization engine does just what you tell it to do, and nothing more.

So if you have an objective to minimize cost, the mathematical optimization engine will return a solution with the lowest cost. If you are looking at the output of this model it may look as though there are "better" solutions that will serve customers better. That is, a customer in Chicago may be getting their product from the warehouse in Dallas and not from the nearby warehouse in Indianapolis. However, you must keep in mind that the optimization engine did not ever consider service in its calculation.

If your objective function includes information about only transportation cost, do not expect the warehouse costs or plant costs to be considered. In fact, a solution based solely on transportation costs may turn out to make bad decisions on the warehouses or plants.

Also, keep in mind that the optimization does not have a mechanism to break ties. To the optimization, a tie is a tie, and whichever solution it finds first, it will return. This can often come up when you have relatively flat transportation costs. That is, it may cost the same to ship to a customer that is 100 miles away as it does to ship to one 500 miles away. So when the mathematical engine chooses to skip the warehouse that is 100 miles away and use the one that is 500 miles away, that is perfectly valid.

Fixing the objective is not always as straightforward as you might think. Sometimes, you can simply add a missing cost element to the objective and the model works fine. Other times, you may not want to add the missing cost, but simply add a constraint to the model to encourage more reasonable solutions. We discussed this earlier in the book when we excluded warehouse costs but put in constraints to limit the number of warehouses it picked.

The reason that fixing the objective is not straightforward is that you sometimes have objectives that cannot simply be added together. For example, you may want to minimize cost and minimize the weighted average distance to customers. We cannot simply combine these. In the section on multi-objective optimization, we talked about the ability to automatically run different objectives at the same time to develop a trade-off curve between the two.

You will find that many real-world problems you will encounter involve multiple objectives. Building a trade-off curve between these objectives is a good way to fix problems related to the objective.

Fixing the Constraints of the Optimization

When setting up the constraints, you need to strike the right balance between giving the model enough constraints to capture the reality of the business and restricting the model too much so that it is very limited in what results it can return.

In the case where there are not enough constraints, the model tends to return results with low costs but unrealistic business results. That is, the model is doing things that are just not possible. In this case, you need to tighten up the model by adding constraints and restricting things that cannot happen in practice. There is a side benefit to not having enough constraints: Sometimes it can allow you to see possible solutions that you didn't think were possible. On more than one occasion, we have seen results from solutions that the client did not think were possible become possible when they saw the potential savings.

In the case where you have overrestricted the model, the results you are getting do not seem to be significantly different from the current state. That is, the model does not have enough freedom to find different solutions. One way this can happen is if you assume that the way things are done now is a constraint and not just a business practice. To avoid this problem, when you are building your model, you want to make sure that current practices are not necessarily listed as a constraint, and you want to make sure that constraints are real constraints. A real constraint is a real physical limit or a legal or contractual issue, or some rule that the company really cannot violate. If you vet your constraints carefully, you can avoid this problem.

Fixing the Decisions of the Optimization

The decisions or choices you allow the model to make are similar to constraints. If you do not give the model many decisions to make, you are limiting potentially different results and are likely seeing very similar answers to the current state of your supply chain. Often it is important to allow the model a variety of decisions to see which ones it takes.

We worked with a client who was looking to open or close warehouses. After many different runs, they found savings ranging from $300,000 to $500,000. Then, they allowed the model to make decisions on which plants to open or close. The savings went up to $7 million. They were not ready to make those kinds of decisions, but the model told them about the opportunity. After about 18 months, they changed the locations of their plants based on the model results.

Sometimes giving the model "too many" decisions can be a good thing that leads to better results.

Fixing the Data of the Optimization

When debugging a model, we cannot forget about the data. Even though this is the most obvious and is where a lot of problems are found, you should not ignore the other elements of an optimization problem.

There are two main problems with the data:

1. It is wrong.

2. It is not accurate enough (go back to our discussion on significant digits).

When we say the data is wrong, we are thinking of possibilities like the following:

1. You have typos—you put in 1,000,000 when you meant 100.

2. You put in 100 when the correct value is 500.

3. When you moved data around, you accidentally assigned demand for Product #1 to Product #2.

4. Because of the way data was pulled, you are missing data for an entire division, product family, or type of customer.

While we can't provide a magic bullet for avoiding the tedious work of checking the data carefully and doing normal data cleaning work (double-checking, looking for high and low values, looking at the top 20%, sorting the data by different criteria—customer, product, region, and so on), we have found that running a baseline and a lot of test scenarios is actually a good way to find problems with the data.

In Chapter 8, we talked about using the current state as a way to validate the model and data. So if the costs match up reasonably well, that is a good way to ensure that you are not missing data or have not misplaced decimal places.

By running the model a lot and forcing certain decisions, you can also look for problems with the data that does not get used in the baseline. That is, when you are optimizing, you may be using plants, warehouses, and transportation moves that are not used in the baseline. By running scenarios, you can see how this data behaves.

Fixing inaccurate data may or may not be an issue. Before you fix inaccurate data, you should review the discussion on significant digits in Chapter 1. In many cases, it may turn out that inaccurate data is actually as accurate as it needs to be. In other cases, it is important to run scenarios in which you change the data in question by +/-10%, +/-20%, and so on, to see how a data element actually changes the decisions. If the model does not change much as you change the data, this tells you not to focus on this data. If the model does change, you can either refine your model or make sure that your final results are robust even if this data does change.

Lessons Learned for the Art of Modeling

In this chapter, we introduced the art of modeling and debugging. Although you use optimization to solve these problems, a lot of thought goes into how you set up these models. This is the art of modeling. This chapter is a good starting point for learning the skills of setting up a good model and properly debugging it. But it takes practice to become a good modeler. If you combine what you learned in the first part of this book—what the finished models look like and how they behave—with the lessons from this chapter, you will be off to a fast start.

End-of-Chapter Questions

1. If you are creating a model and one of the key considerations is which port of entry you should use, why would you model the port like you would model a warehouse? What costs would be relevant for picking which port to use?

2. You are creating a model for a manufacturing company with four major plants and ten warehouses. They want to reduce costs by determining the optimal number and location of the plants and warehouses. During the initial set of meetings, one of the team members is very concerned about modeling the "samples" that the firm sends out. As part of the marketing of the product, the firm sends out quite a bit of sample product. This product is pulled from the plants or warehouses and shipped to customers via small-parcel. It accounts for about 10% of the demand records (and about 0.5% of the demand volume) and about 3% of the logistics costs. Should you include this part of the business in the model?

3. Open the model found in the file `Debugging Example I.zip` on the book Web site. In this model, the customer wants to minimize the average distance to customers. However, when the model runs, one warehouse serves the majority of the customers even though there are other warehouses that could be used and are closer to the customers. Why is this happening? How would you fix this model to remedy the situation?

4. Open the model found in the file `Debugging Example II.zip` on the book Web site. This model will not solve. Determine what is preventing the model from solving and how you would fix the problem.

5. Open the model found in the file `Debugging Example III.zip` on the book Web site. This model solves, but is returning a cost that is about 100 times higher than what you expect the result to be. Determine what the problem is and suggest how you would remedy the situation.

6. You are starting a network modeling project. When you ask the project manager what the objective of the study is, you get the answer, "We want to optimize the entire end-to-end supply chain." Why doesn't this answer help you? What additional information do you need?

7. Why shouldn't your model be an exact one-to-one replica of the supply chain?

8. Using the diagram of the retailer from this chapter, draw the diagram of a retailer in which the urban stores are served only from regional warehouses, the rural stores are served from either the central warehouse or regional warehouses, and the online customers are served from a standalone warehouse that receives product directly from the vendors.

13

DATA AGGREGATION IN NETWORK DESIGN

In this chapter, we will focus our attention on one of the most interesting, yet challenging parts of network design—data aggregation (we'll refer to it as just "aggregation"). Aggregation is putting the data of your supply chain into logical groups for the purposes of modeling. For example, instead of modeling every single item that moves through the supply chain, our model may contain only a small handful of product groups. When we aggregate, we typically aggregate the following:

- Products (from many different individual products to a small number of product families)

- Customers (from many different actual delivery locations to several hundred aggregate points representing all the customers in a geographic area with other similar traits)

- Plants, warehouses (that are a different building but on the same campus), and vendors (from hundreds of vendors to a small group of important vendors or vendors in the same geographic area)

- Time periods (modeling a year instead of every single day)

- Cost types (from many different line items to a single cost figure)

One of the reasons that aggregation is interesting is that it is both natural in how we think about problems and scary because we worry that we aren't being accurate.

Throughout this book, we have used aggregation strategies without calling them out. Al's example grouped all types of products into one weight based product group; the small-parcel example looked at shipments to about 200 customer points representing 10,000 or more actual shipment points; and so on. In each of these cases, we thought of the models as good representations of the supply chains. They seemed natural. Also, when we diagram a supply chain on a white board, we think of elements of our supply chain in groups. For example, we can draw flows for products that travel in full pallets and flows for products that travel in bulk containers. Again, all very natural.

However, when it comes to actually aggregating the data to make decisions, we lose our nerve, forget what is natural, and get scared thinking that if we don't directly load all 10,000 products and 45,000 shipment points into the model, it won't be accurate.

We need to get over our fear and learn the techniques of aggregation. Only firms that sell a very small number of products or ship to a small number of endpoints can avoid it. And even these firms may want to aggregate.

First, there are some technical reasons for why you want to aggregate:

- The optimization models we've covered in this book can be difficult to solve. If you simply loaded every product, every vendor, and every delivery point, you will likely find yourself with an optimization model that does not fit into the memory of the largest computer and has no chance of solving.

- If you are running a model for future decisions (and especially those more than several months or a year out), your forecasts at individual products or sites are not close to accurate.

On top of the fact that you technically can't avoid aggregation (which, sadly, won't be enough to convince some people), there are some very practical reasons you want to aggregate:

- Cost and time involved in obtaining, processing, validating, and analyzing data at a very detailed level is prohibitive.

- It can be impossible to understand the big-picture model and the influence and interaction between the key data elements if you are working at a detailed level.

- When the final decisions are made, the executives are going to go back to the natural way of thinking about the supply chain and think about the model in aggregate terms.

Because you cannot avoid aggregation, don't fear it. When aggregation is performed well, models will closely represent the real world without sacrificing accuracy, but at the same time provide manageable and understandable outputs that lead to sound decisions.

Aggregation is also interesting because there is an art to it. The lessons we learned in Chapter 12, "The Art of Modeling," also apply to aggregation. So you will need to use your creativity to solve problems. But there are also some best practices and science we can apply to make our aggregation decisions better. We will cover this later in the chapter.

Finally, this is interesting because there are many ways to approach aggregation and they may be different in each model or project. And the same business may use different aggregation strategies to answer different types of questions.

It is important to start thinking about aggregation early in the project—as early as the first discussions about the model. Following are some key questions to ask and consider to help drive the right strategies:

- What exactly are we looking to solve for, and what portions of the supply chain will we realistically be able to change? By answering this question, you can get a feel for which data elements are more or less important. If a data element is less important, it is a good candidate for aggressive aggregation. If you are answering questions about plant locations relative to warehouses that won't change, it might make sense to roll all your customers to the warehouse that serves them.

- What data is available? If there are areas where you don't have good data, you might as well aggregate.

- Are there detailed constraints to model? In general, the more detail you want to model in the constraints, the less aggressive you can be in aggregation. If you want to build constraints for just your top customers, you have to split your customers in a geographic area into at least two groups (one for the top and one for the others).

- How are the transportation rates applied? If we are looking at simple transportation structures, it may make no sense to add extra details. For example, if we are looking at minimizing the delivery cost of products from our warehouse to customers and all products travel in full pallets, then we might not need any detail on the products.

In the next few sections, we will introduce aggregation strategies for each of the elements of the model as well as discuss each of these strategies in more detail.

Aggregation of Customers

Customers are one of the most commonly aggregated elements in network design models. As a reminder, we are using the term "customer" to refer to ship-to locations, so if you ship product to a firm with ten locations, we are talking about the ten locations. A typical supply chain will include hundreds or thousands of customers to which products are distributed and where they finally get consumed. These sites may even change from year to year as the customer base shifts. Customer aggregation usually comes down to geographic proximity and types of customers.

It is pretty clear that customers that are physically close to each other should be grouped together. That is, if you have 25 customers in Denver, you might consider grouping these customers together. Then, if you have 3 distinct types of customers, you may split the Denver customer group into 3. You've gone from 25 points to 3 and still have the essence of your problem.

The most common strategy for grouping customers by geography is to use some form of the ZIP or postal code, or a metro area, or regional area.

For U.S.-based analyses, there are too many five-digit ZIP Codes, so it is most common to aggregate customers by three-digit ZIP Codes. (i.e., the first three digits of the standard five-digit ZIP Code). To pick the point to plot on the map, it is most common to use the largest single customer or five-digit ZIP Code. In general, there tends to be more three-digit ZIP Codes in largely populated areas and fewer in sparsely populated areas. The maps in Figure 13.1 show examples from Illinois and Montana.

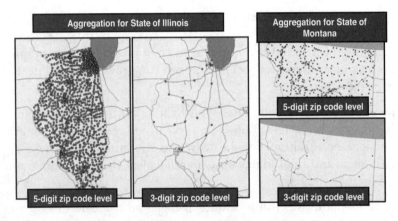

Figure 13.1   Example Showing Aggregation at Five- and Three-Digit ZIP Level for Illinois and Montana

In the U.S., most transportation rate structures—especially small-parcel and LTL, are defined or estimated at the three-digit ZIP Code level. So, in one sense, you lose no accuracy by grouping by three-digit ZIP Codes.

In Europe, two-digit postal codes can be used like three-digit ZIP Codes in the U.S. Then, in all regions of the world, cities and regions can be good geographic zones. When doing worldwide studies, you may even aggregate all the customers in one country to a single point. Or if you have a model that is mostly focused on the U.S., you may aggregate your non-U.S. customers to the port of exit. The actual method chosen largely depends on the quality of data and how well it ties to the transportation rate structure.

In some countries, aggregation to the district or county level may make sense if aggregation by city does not yield a significant reduction in the number of customers. Figure 13.2 shows the difference in aggregation at city versus district level for locations in India.

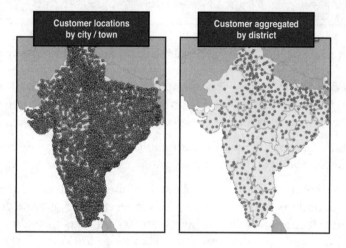

Figure 13.2   Example Showing Aggregation of Locations in India
at City/Town and District Levels

For all models, it is very common to remove outlier and low-demand locations with insignificant demand that will always be difficult to service. For example, Alaska, Hawaii, and Puerto Rico in the U.S. are either modeled at the nearest continental U.S. port or excluded from the model. Or simply removing the low-volume demand points within your data may also save you time and effort.

In addition to geographic strategy, customers can be classified and aggregated using other key categories. You can think of this as stacking several points on top of the geographic aggregation. For example, if you aggregate by three-digit ZIP Code, you may have two points at each three-digit ZIP Code representing the two types of customers you want to model. Some examples of categories include these:

- **Required service levels**—For example, next-day customers and three-day customers. This provides the ability to identify and model different service levels and see the impact on the network strategy.

- **Shipping methods**—For example, LTL, TL, Rail. This allows the user to easily apply transportation rates and business rules by classifying customers that are served by LTL shipments versus TL shipments. See Figure 13.3 as an example.

Customer Name	City	State	Total Wt by Mode			% Wt by Mode		
			Truckload	LTL	Parcel	Truckload	LTL	Parcel
Customer 1	Norman	OK	243,535	45,464	4,354	83%	15%	1%
Customer 2	Ft. Lauderdale	FL	75,645	4,643,546	476,464	1%	89%	9%
Customer 343	Des Moines	IA	3,235,325	5,262,352	3,253,252	28%	45%	28%
Customer 445	Carson City	NV	43,224	24,242	252,552	14%	8%	79%
Customer 134	Bridgetown	CT	76,433	63,536	767,465	8%	7%	85%

Figure 13.3   Example Showing Breakdown of Volume by Customer by Mode

- **Type of delivery location**—For example, a store or an online customer. This provides the ability to track and apply specific business rules based on the type of end consumption point for the products.

You can be creative in coming up with customer categories. Just be sure you don't create so many categories that you have more customers than in your original data set.

Also, don't forget about the art of modeling and coming up with creative solutions. For example, there is nothing that says that when you decide to aggregate customers you have to aggregate all customers. If you have 20 to 30 customers that account for a large portion of your demand, you may simply model those customers as they are and aggregate all the others.

Validating the Customer Aggregation Strategy— National Example

We may need to validate the accuracy of our aggregation strategy before others in the organization will accept it. We'll do two tests, one at a national level and one more detailed for a region. You may need to come up with other ways to validate your aggregation strategy.

This sample model has the following data elements:

- One product sourced from one plant.

- Our objective is to minimize transportation costs only.

- Consider two sets of potential warehouses:

   1. 26 potential warehouses—major U.S. distribution points.

   2. 60 potential warehouses—major U.S. cities/metropolitan statistical areas.

- For customer data we will use U.S. population data obtained from the U.S. Census Bureau Web site (http://factfinder2.census.gov/faces/nav/jsf/pages/index.xhtml). We will model this data at two levels of aggregation with this model:

  1. Top 16,000 five-digit ZIP Codes with a population of at least 3,000 residents. We will exclude ZIP Codes in Alaska, Hawaii, and Puerto Rico for this example.

  2. ZIP Codes aggregated to three-digit ZIP Code level. This yields a total of 876 three-digit ZIP Codes. For the model, we will pick the largest five-digit ZIP within each three-digit ZIP for geocoding.

- The total demand is the same in both cases.

For the first model, we will load data for 16,000 customers and 60 potential warehouses. For the aggregated model, we will load the model with the 876 three-digit ZIP Codes and the top 26 cities as potential warehouse locations.

The charts in Figure 13.4 give us a graphical view of the inputs for the two customer aggregation strategies. We can see the significant difference in density of customers between the two models. However, the real question is whether this increased level of detail offers any significant benefits in terms of the quality of outputs and associated decision making.

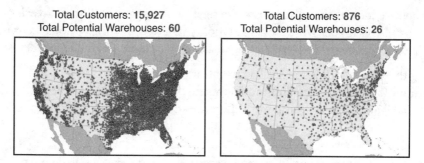

Figure 13.4 Comparison of Input Data—16,000 Versus 800 Customers

For purposes of simplicity, we are using a truckload rate of $2/mile and 40,000 units per load.

To test the models, we will run a scenario on each model to pick the best five warehouse locations from the given list of potential sites. The results of the two models are shown in Figure 13.5.

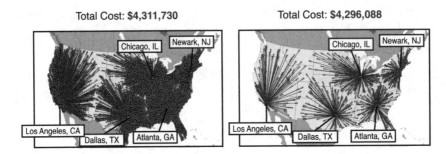

Figure 13.5   Comparison of Output Results

The charts themselves give insight into how the models compare. Both models picked the same five warehouses: Newark, New Jersey; Atlanta, Georgia; Dallas, Texas; Chicago, Illinois; and Los Angeles, California. This shows that with the right aggregation strategy, we can come up with the same answer as one without aggregation. The model with three-digit ZIP Codes is much easier to run and analyze compared to the unaggregated model.

Now when we look at the actual costs, we see that the aggregated model varies by only 0.363% from the model without aggregation. As noted in the previous example, it is likely that the forecasted demand has more error than the difference in these solutions.

By using three-digit ZIP Code level aggregation, our customer file went from 16,000 to 800 without loss of accuracy. This model was further simplified by reducing the number of potential warehouses, without any difference in the final solution recommendation.

Validating Customer Aggregation—Regional Example

In the previous example, we analyzed and validated the impact of customer aggregation on a broad, national level. In this example, we will focus on a specific market (Chicago) and see what kind of impact aggregation may have. This gives you a deeper look at the impact of aggregation.

The map in Figure 13.6 shows a set of 30 ship-to locations in the Chicago area, at the five-digit ZIP Code level. Each customer location receives between 5 and 20 full truck-load shipments a week.

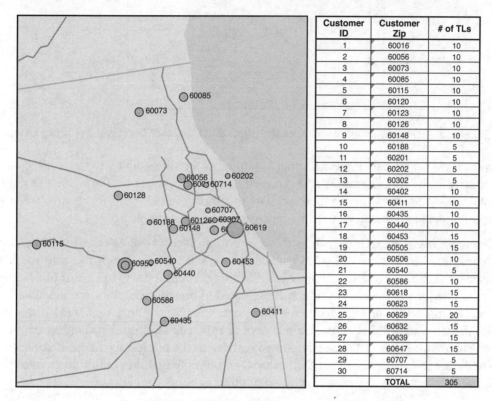

Customer ID	Customer Zip	# of TLs
1	60016	10
2	60056	10
3	60073	10
4	60085	10
5	60115	10
6	60120	10
7	60123	10
8	60126	10
9	60148	10
10	60188	5
11	60201	5
12	60202	5
13	60302	5
14	60402	10
15	60411	10
16	60435	10
17	60440	10
18	60453	15
19	60505	15
20	60506	10
21	60540	5
22	60586	10
23	60618	15
24	60623	15
25	60629	20
26	60632	15
27	60639	15
28	60647	15
29	60707	5
30	60714	5
TOTAL		305

Figure 13.6 Customer Locations in Chicago Area, and Weekly Demand in Number of Truckloads

For this validation exercise, we will estimate the transportation costs to serve this weekly demand from a set of warehouse locations in Indianapolis, Indiana; New York, New York; Atlanta, Georgia; and Chicago, Illinois (see Figure 13.7). We will estimate the costs if these customers were to be served from each of these warehouses as potential options.

Figure 13.7   Map Showing Warehouse Locations and Customer Locations in Chicago Area

We will then aggregate these customer locations at the three-digit ZIP Code level, and recalculate the transportation costs from the same warehouse locations for the same total weekly demand. We can then compare the transportation costs for the two scenarios to understand the impact of this aggregation strategy.

Before we begin, let us look at how the transportation costs will be calculated. Because these customers are served by full truckloads, we will use a cost-per-mile rate applied against the distance to estimate the costs. For this example, we will use a cost of $2/mile as the rate for the truckload shipments. As noted in previous chapters, truckload carriers also apply a minimum charge for shipments that fall within a very short distance. For this case, we will use a minimum charge of $400 per TL shipment. This means that we will use the larger of the distance-rated cost versus the minimum charge. Because the warehouse in Chicago is one of the options, we can note that the minimum charge will apply because the distances fall within 200 miles.

The table in Figure 13.8 shows the transportation cost estimates by customer location (five-digit ZIP Code). We can see that the costs from Chicago to all customer locations are based on the minimum charge because the distances are less than 200 miles. For example, the cost for customer ID 1 (60016) is $400 * 10 truckloads, or $4,000. We also see that there are a few ZIP Codes that are less than 200 miles from Indianapolis and hence the minimum charge is being applied.

Customer ID	Customer Zip	Number of Truckloads	Origin Warehouse							
			Atlanta		Chicago		Indianapolis		New York City	
			Distance (miles)	Transportation Cost	Distance (miles)	Transportation Cost	Distance (miles)	Transportation Cost	Distance (miles)	Transportation Cost
1	60016	10	703	$ 14,057	21	$ 4,000	214	$ 4,286	855	$ 17,107
2	60056	10	706	$ 14,115	24	$ 4,000	217	$ 4,346	857	$ 17,145
3	60073	10	731	$ 14,618	48	$ 4,000	242	$ 4,849	869	$ 17,387
4	60085	10	728	$ 14,564	42	$ 4,000	237	$ 4,744	854	$ 17,090
5	60115	10	712	$ 14,234	68	$ 4,000	237	$ 4,741	906	$ 18,130
6	60120	10	711	$ 14,220	42	$ 4,000	277	$ 4,544	878	$ 17,566
7	60123	10	711	$ 14,220	42	$ 4,000	227	$ 4,544	878	$ 17,566
8	60126	10	693	$ 13,859	19	$ 4,000	206	$ 4,120	857	$ 17,147
9	60148	10	693	$ 13,858	23	$ 4,000	207	$ 4,141	862	$ 17,231
10	60188	5	699	$ 6,986	31	$ 2,000	214	$ 2,139	869	$ 8,689
11	60201	5	699	$ 6,992	13	$ 2,000	208	$ 2,084	842	$ 8,421
12	60202	5	699	$ 6,992	13	$ 2,000	208	$ 2,084	842	$ 8,421
13	60302	5	689	$ 6,889	9	$ 2,000	200	$ 2,001	848	$ 8,476
14	60402	10	686	$ 13,720	11	$ 4,000	198	$ 4,000	848	$ 16,963
15	60411	10	657	$ 13,133	31	$ 4,000	169	$ 4,000	838	$ 16,763
16	60435	10	669	$ 13,376	42	$ 4,000	189	$ 4,000	868	$ 17,361
17	60440	10	681	$ 13,626	31	$ 4,000	198	$ 4,000	865	$ 17,300
18	60453	10	675	$ 13,504	16	$ 4,000	187	$ 4,000	846	$ 16,911
19	60505	15	691	$ 20,725	42	$ 6,000	210	$ 6,311	878	$ 26,352
20	60506	10	691	$ 13,817	42	$ 4,000	210	$ 4,208	878	$ 17,568
21	60540	5	688	$ 6,876	33	$ 2,000	205	$ 2,050	870	$ 8,699
22	60586	10	677	$ 13,549	41	$ 4,000	198	$ 4,000	873	$ 17,458
23	60618	15	683	$ 20,491	5	$ 6,000	193	$ 6,000	841	$ 25,244
24	60623	15	683	$ 20,491	5	$ 6,000	193	$ 6,000	841	$ 25,244
25	60629	20	683	$ 27,321	5	$ 8,000	193	$ 8,000	841	$ 33,658
26	60632	15	683	$ 20,491	5	$ 6,000	193	$ 6,000	841	$ 25,244
27	60639	15	683	$ 20,491	5	$ 6,000	193	$ 6,000	841	$ 25,244
28	60647	15	683	$ 20,491	5	$ 6,000	193	$ 6,000	841	$ 25,244
29	60707	5	693	$ 6,926	12	$ 2,000	204	$ 2,038	850	$ 8,496
30	60714	5	700	$ 7,000	16	$ 2,000	210	$ 2,105	849	$ 8,492
	TOTAL	305		$ 421,633		$ 122,000		$ 127,332		$ 522,615

Figure 13.8   Transportation Cost Estimates for Customers at Five-Digit ZIP Code Level

Now, we will aggregate the customer locations to the 3-digit ZIP Code level for this exercise. The chart in Figure 13.9 shows the demand calculation with aggregating from 5-digit ZIP to 3-digit ZIP level. The total number of truckloads is maintained at 305 loads. We will use the largest 5-digit ZIP within each 3-digit ZIP for geo-coding in the model. So, the number of customer points in the model has decreased from 30 to 8.

Customer ID	Customer Zip	# of Truckloads	3 digit Zip
1	60016	10	600
2	60056	10	600
3	60073	10	600
4	60085	10	600
5	60115	10	601
6	60120	10	601
7	60123	10	601
8	60126	10	601
9	60148	10	601
10	60188	5	601
11	60201	5	602
12	60202	5	602
13	60302	5	603
14	60402	10	604
15	60411	10	604
16	60435	10	604
17	60440	10	604
18	60453	10	604

Customer ID	Customer Zip	# of Truckloads	3 digit Zip
19	60505	15	605
20	60506	10	605
21	60540	5	605
22	60586	10	605
23	60618	15	606
24	60623	15	606
25	60629	20	606
26	60632	15	606
27	60639	15	606
28	60647	15	606
29	60707	5	607
30	60714	5	607

Total		305	

3 digit Zip code	# of Truckloads	5-digit zip for geocoding
600	40	60085
601	55	60148
602	10	60201
603	5	60032
604	50	60402
605	40	60505
606	95	60629
607	10	60707
TOTAL	305	

Figure 13.9   Aggregation from Five-Digit to Three-Digit ZIP Code Level

We will use the same rate structure for this model including the minimum charge. The results of the analysis at the three-digit ZIP level are shown in Figure 13.10.

Customer Zip	Number of Truckloads	Atlanta		Chicago		Indianapolis		New York City	
		Distance (miles)	Transportation Cost	Distance (miles)	Transportation Cost	Distance (miles)	Transportation Cost	Distance (miles)	Transportation Cost
600	40	728	$ 58,254	42	$ 16,000	237	$ 18,976	854	$ 63,360
601	55	693	$ 76,220	23	$ 22,000	207	$ 22,775	862	$ 94,772
602	10	699	$ 13,985	13	$ 4,000	208	$ 4,167	842	$ 16,841
603	5	689	$ 6,889	9	$ 2,000	200	$ 2,001	848	$ 8,476
604	50	686	$ 68,601	11	$ 20,000	198	$ 20,000	848	$ 84,816
605	40	691	$ 55,267	42	$ 16,000	210	$ 16,830	878	$ 70,273
606	95	683	$ 129,776	5	$ 38,000	193	$ 38,000	841	$ 159,878
607	10	693	$ 13,851	12	$ 4,000	204	$ 4,076	850	$ 16,991
Grand Total	305		$ 422,844		$ 122,000		$ 126,824		$ 520,406
Results with 5-digit ZIP Code			$ 421,633		$ 122,000		$ 127,332		$ 522,615
% Difference			-0.29%		0.00%		0.40%		0.42%

Figure 13.10   Transportation Cost Estimates for Customers at Three-Digit ZIP Code Level

The results show that the costs from each warehouse for the three-digit model are very close to that of the five-digit ZIP model. The differences are in the range of –0.29% to +0.40%, which is certainly smaller than the error in the demand forecasts or in likely transportation rate changes because of changing oil prices.

We also notice that there is no difference in costs from the Chicago warehouse as the minimum charge applies again for all locations. The differences in distances due to aggregation are eliminated due to the minimum charge. This is a key point to take note when aggregating at a regional level. Even though the five-digit ZIP level model shows a greater level of detail in actual locations, this does not offer any additional advantages in terms of costs.

To summarize, this validation exercise helped us understand how the right aggregation strategy can still provide valid and accurate results even at the regional or subregional level.

Aggregation of Products

Product aggregation tends to be more challenging than customer aggregation. With customer aggregation, it is hard to go too wrong grouping customers by proximity. With product aggregation, most firms already have products grouped into marketing families. Unfortunately, there is nothing about a marketing family that naturally makes it a good candidate for an aggregate product group for a supply chain model. For example, a retailer may have a marketing family for Women's and Men's Apparel. From a logistics viewpoint, it might be more natural to have the socks be in one product family because they may come from the same vendor, have the same relative size, and require the same shipping and handling methods. On the other hand, the hanging clothes may naturally fit together because they are shipped and handled together.

So products sharing logistics characteristics make for good product families in a network design model. The marketing families are often irrelevant.

This means that you, the supply chain team, may need to create new product families and have the people in the organization agree to their validity.

Before we go too deep into product aggregation, there are two key questions you should ask at the start of every project. These questions will save you a lot of time in product aggregation and may help you avoid complicated product aggregation exercises.

The first key question is this: *Can you create one aggregate product representing everything?*

If you think about this from a marketing point of view, you will never answer "yes" to this question. However, from a logistics point of view, if you care only about pounds, kilos, pallets, cartons, or boxes moving through the supply chain, you should use it. In the examples in this book, we showed legitimate models that relied just on total weight or total packages sent through the supply chain, with no details on the underlying products. In practice, many good models are built with one product family.

After thinking long and hard about the first question, the second critical question is this: *If we can't create one aggregate product family, will two do?*

Of course, this question is really many questions, because you may want to ask about three, four, and so on. You can see that the preceding questions suggest a way of thinking about aggregation that will help you. We have found, in practice, that it can be easier to start your product aggregation with one aggregate product (say pounds) and then grudgingly add more products as needed. This is opposed to starting with all 10,000 individual products and trying to whittle it down. In the latter case, it is always harder to get to a manageable number of product families.

As you start to add more product groups, or your hand is forced, and you must start with all products and whittle it down, you need to know what characteristics to consider. What is important is that you consider the logistics characteristics of the products. The following sections provide some guidelines for how to group products.

Consider the Source of the Products

The source of the products is often the most overlooked category and can cause serious flaws in your model if you do not properly account for it. When including the source of your products, you are grouping products by where they came from. This helps ensure the correct flow from the plants to the final customers.

As an example, if you are building a model for the supply chain of apples, you could start by grouping all types and sizes of apples into one product called "apples." However, if your red apples come from Michigan and yellow apples from Washington, you need to include this in the aggregation strategy. Customers will demand both red and yellow apples. If you group products into just "apples," the customers close to Michigan will get all their apples from the Michigan orchards and the ones on the West Coast will get all their apples from Washington. There would be nothing in the model to force customers to get apples from both Michigan and Washington. To make this happen, you need to create a product grouping that includes the source. For example, the table in Figure 13.11 shows some of the types of apples in this example.

SKU ID	Source Vendor	Product Group
SKU 1	Michigan	Michigan apples
SKU 2	Michigan	Michigan apples
SKU 15	Michigan	Michigan apples
SKU 53	Washington	Washington apples
SKU 68	Washington	Washington apples
SKU 80	Washington	Washington apples

Figure 13.11   Example of Apple SKUs Categorized by Source Vendor Location

The correct way to model this would be to aggregate the three SKUs from Michigan as "Red Apples—Michigan sourced," and the remaining three SKUs as "Green Apples—Washington sourced," and allow only Michigan apples to come out of Michigan and the same for Washington. This will now correctly reflect the flow of products through the supply chain.

In more complex situations, a general way to think about categorizing products by sources is to think of sourcing groups. A sourcing group consists of all the SKUs or products that are made from the same set of plants. This structure allows you to capture the fact that products can be made in multiple locations. As an example, the following table shows how you might go about this task.

The table in Figure 13.12 shows that SKU 1 and SKU 8 are sourced from Plant 1, and therefore placed in Group 1. SKU 7 and SKU 12 are both sourced from Plants 1 and 2, and therefore placed in Group 1-2.

SKU	Source Plants	Source Group
1	Plant 1	Group 1
2	Plant 2	Group 2
3	Plant 3	Group 3
4	Plants 1,2	Group 1-2
5	Plants 1,2,3	Group 1-2-3
6	Plant 2	Group 2
7	Plant 1,2	Group 1-2
8	Plant 1	Group 1
9	Plant 2	Group 2
10	Plant 2	Group 2
11	Plants 1,2,3	Group 1-2-3
12	Plants 1,2	Group 1-2
13	Plants 1,2,3	Group 1-2-3
14	Plants 1,2	Group 1-2
15	Plant 2	Group 2

Figure 13.12   Categorization of SKUs into Source Groups

Consider Removing Products with Low Volumes

Part of the art of modeling is to separate the significant from the trivial. And, often, when a firm makes or sells thousands of products or SKUs, they will have many very low-volume products. The chart in Figure 13.13 shows a typical example of the number of products sorted from highest demand to lowest and showing the cumulative percentage of demand.

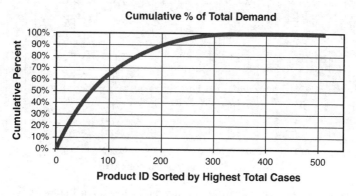

Figure 13.13   Product Demand Pareto Chart

When we analyze this, we immediately find what most firms find with their data: 90% of the demand flowing through the network is made of less than half of the total 500 products. Taking that further, we see that 99% of the demand is made up of only 100 more products, for a total of 300 products. So, 300 of the 500 products make up 99% of the demand, and we can easily remove the remaining 200 products from our analysis.

Consider the Size of the Products

The size of the products can impact transportation costs and warehousing costs. So products of similar size could be grouped together.

A perfectly valid and simple way to do this is to group products by small, medium, and large. As an example, in reality, this translates into products shipped in boxes, those on pallets, and bulk items.

A more sophisticated way is to cluster the products by their weight and cube (or cubic volume). This will help accurately model transportation costs while reducing the complexity of the model. The graph in Figure 13.14 shows an example of clustering.

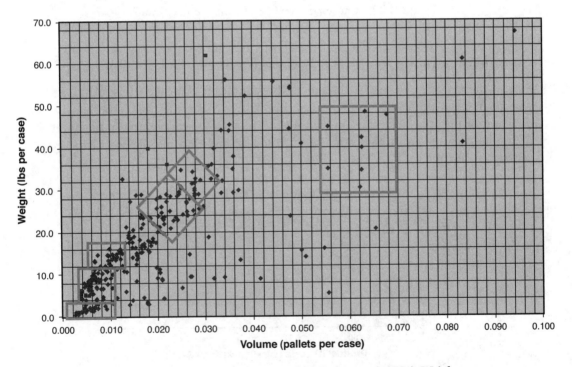

Figure 13.14 Plot of Products and Their Respective Unit Weight

The graph shows a plot of the products based on their unit weight (lbs per case) and volume (pallets per case). The rectangles shown on the graph illustrate how to cluster SKUs based on similar weight and cube characteristics. There are sophisticated ways you can do this type of cluster analysis that are beyond the scope of this book. However, keep in mind that you want to try to keep your aggregation strategy as simple as possible.

Consider Different Packaging Requirements

It is also common to aggregate products based on their packaging requirements that may help easily model manufacturing and transportation costs. For example, the products for a beer manufacturer can be classified into bottles, cans, and kegs (see Figure 13.15). We can further classify bottles into 6-pack, 12-pack, and 24-pack, allowing the user to model specific constraints and rules for each package type.

SKU ID	SKU Name	Package
4324	ABC 12oz 12pk can	12pk
4141	ABC 12oz 24pk bottle	24pk
2553	AB4lime 16oz 6pk bottle	6pk
3652	AB4 24oz 12pk bottle	12pk
7676	OHJ 12oz 6pk can	6pk
6546	OGH 0.5bbl keg	-
3645	OGH 24oz 6pk can	6pk

Figure 13.15   Example of Beverage SKUs with Package Attribute

Consider Production Requirements

The same criteria can be applied to production requirements in order to simplify sourcing and production constraints and rules. For example, products that require special coating can be classified and tagged, allowing these products to be easily assigned to the specific production lines in the model.

Consider Products That Share Components or Raw Materials

Products that share the same components or raw materials can be aggregated together in order to simplify the modeling of bills-of-material and associated business rules for these products. A model for a chewing gum manufacturer would classify products as sugar and sugar-free based on products that use sugar or sugar substitute, allowing the modeling of sourcing rules.

Consider Products That Share Transportation Requirements

Products that require special transportation modes or requirements can be aggregated together to allow modeling of transportation rules. For example, products that ship in liquid form would be separate from the same product that may be put in a barrel first.

Consider Per-Unit Production Costs

In some supply chains, there may be several products within the same product family or group or across product groups that have significant variation in the unit production costs. This may be attributed to variations in some component or raw material, or the differences in the production process itself. Depending on how these other portions (raw material and BOM, production lines) are modeled, it may be easier to group the SKUs based on the unit production costs to simplify the modeling.

Consider Predefined Product Families

In some cases, if the company has experience with network modeling, the existing product families may work. Of course, if the firm has experience with network modeling, they likely will have gone through the steps listed previously.

For example, a retailer may classify its SKUs into frozen goods, dry goods, fresh produce, liquid beverages, and so forth. Or a beverage producer may group products as shown in the table in Figure 13.16.

SKU ID	SKU Name	Type	Size	Brand	Package
4324	ABC 12oz 12pk can	Can	12oz	ABC	12pk
4141	ABC 12oz 24pk bottle	Bottle	12oz	ABC	24pk
2553	AB4lime 16oz 6pk bottle	Bottle	16oz	AB4	6pk
3652	AB4 24oz 12pk bottle	Bottle	24oz	AB4	12pk
7676	OHJ 12oz 6pk can	Can	12oz	OGJ	6pk
6546	OGH 0.5bbl keg	Keg	0.5bbl	OGH	-
3645	OGH 24oz 6pk can	Can	24oz	OGH	6pk

Figure 13.16   Example of Beverage Products with Key Attributes Useful in Aggregation

If this information is readily available, you should use it.

Testing the Product Aggregation Strategy

As a next step, we will look at testing an aggregation strategy in a model with some sample data. For this test exercise, we will build a sample model with the following elements:

- 5 plants

- 25 potential warehouses

- Distance-based service constraints

- Inventory holding costs

- Fixed warehouse costs

- Product aggregation—we will test the model with the following options:

  - 46 original SKUs

  - 4 aggregated products—aggregated products were created using weighted averages and clustering (similar to the methodology used in a previous example)

The model was run with 46 products, and then with 4 aggregated products. The results of the two scenarios are shown in Figure 13.17. The analysis shows that the aggregated model was 0.0003% higher than the one without aggregation. The difference is that one model picked Seattle, Washington to cover the northern part of the West Coast and the second model picked a location in Northern California.

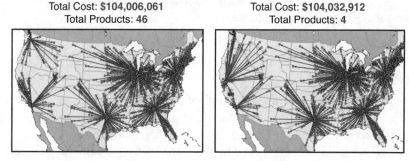

Total Cost: $104,006,061
Total Products: 46

Total Cost: $104,032,912
Total Products: 4

Figure 13.17   Results Comparison for 46 Original Products Versus 4 Aggregated Products

So the critical question is, Does the Seattle solution really differ from the Northern California solution? To understand this question, we will run two scenarios with each model: one scenario forcing Seattle and another scenario forcing Northern California. The results are shown in Figure 13.18.

Aggregation Strategy	Seattle or N. Cali	Total Cost	Difference from Optimal 46 Products
None; 46 Products	Seattle	$104,006.061	0.00%
None; 46 Products	N California	$104,113,011	0.10%
4 Products	Seattle	$104,087,045	0.08%
4 Products	N California	$104,032,912	0.03%

Figure 13.18   Comparison of Results with Seattle and Northern California

If we look at the results for the model without aggregation, we see that there is a difference of 0.1% between the Seattle and the Northern California solutions. This difference is extremely small considering the margin of error and level of accuracy of the underlying data within the model. It is likely that the forecasted demand data has more error than the difference shown here. This helps us conclude that the Northern California solution is equivalent to the Seattle solution based on the cost factors analyzed in the model. This means that the solution requires further analysis evaluating other factors such as labor availability and logistics infrastructure.

This also shows that the aggregated model is still accurate given the margin of accuracy for normal models. The key take-away from this exercise is that with the right aggregation strategy, we can get the same level of accuracy in decision making as one would achieve with modeling without aggregation.

Aggregation of Sites

In most models, you have a limited number of plants and warehouses. So you usually keep these as individual sites and do not need to aggregate. The one minor exception is when you may have two to three plants or two to three warehouses that are all on the same campus or within the same city. In this case, as long as the facilities do similar things, you may group them together.

This is not always the case for vendors. Many firms have hundreds and maybe thousands of vendors.

If we need to include vendors in our model, we treat this a lot like we did with customers. We group vendors by geographic proximity and maybe by type. The example in Figure 13.19 shows vendors grouped by three-digit ZIP Code in Georgia, but we can often build good models by grouping vendors by a larger geographic area like a U.S. state, country, or even a port of entry . As another example, in a U.S.-focused model you may aggregate all your Asian vendors to a single point in Long Beach, California.

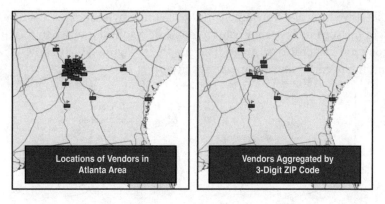

Figure 13.19   Example Showing Aggregation of Vendors in Atlanta Area

Aggregation of Time Periods

When aggregating time periods, we are simply determining the time buckets we will use for our model.

The industry standard for time-period buckets in network design models is a year. Within a year, you capture the full range of demand across seasons, and it is natural to think of reporting your costs and savings in terms of annual numbers. And the nature of most decisions about when to open or close a facility does not need more granularity than an annual bucket.

With this structure, nothing prevents you from running what-if scenarios with different demand patterns. That is, you can run a model with the expected demands five and ten years out. Even more so than with products, we want to make sure we are careful when we model more than a single annual time period.

There are two basic problems we solve when we include multiple time periods in the same model. The first problem we can address is to determine what year to open and close facilities. To capture this, we could model multiple annual buckets in the same model and have the model pick which year we open or close facilities. Note that this makes the model much more complicated. One way to avoid this complication is to simply run different what-ifs with future demand scenarios. You may think you are losing accuracy, but in reality, predicting demand several years out is very difficult anyway, so you may be just introducing false precision by including multiple years in the same model.

The second problem we can address is to determine how to handle seasonality. In this case, we may build models that have 4 quarterly buckets to make different decisions in different parts of the year. Or if the prebuilding of inventory is important, we may model in 12 monthly buckets.

For example, if we look at a consumer products company that experiences a high spike in demand in the summer, we may want to make decisions about when to start building the inventory and where to store it. The graph in Figure 13.20 shows the spike in demand and the need for extra storage capacity.

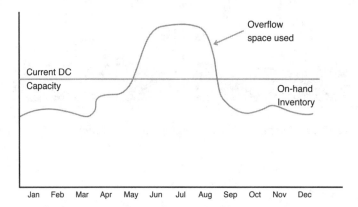

Figure 13.20   On-Hand Inventory Versus Warehouse Capacity

These are some general rules for thinking about time periods. As with all modeling, you can get creative to solve your specific problem. For example, we have seen other cases in which a firm runs a network model with weekly buckets to decide what product is made where and shipped to which warehouse.

In general, as you add time periods to the model, you need to remove other types of decisions. For example, if you have monthly buckets, you are typically fixing the locations of your plants and your warehouses (except the overflow warehouses).

Aggregation of Cost Types

Although cost types may not be commonly thought of as an aggregation decision, you also want to simplify your modeling by grouping similar cost types together.

For example, you may use 100 different full-truckload carriers. Instead of trying to capture each of these, you may want to group them all together and use the average of their rates when they share a source and destination combination. This makes modeling easier and it avoids the false precision of forecasting next year's rates for each of 100 different carriers.

The same lesson applies to costs in your facilities. You may have different components of fixed and variable costs at your plants and warehouses. In fact, you may have many different workers each with a different salary and benefits package. You do not want to model all that detail. It is just as accurate to roll these costs up to a few small categories.

Lessons Learned on Aggregation

You will not be able to avoid aggregation. But you also shouldn't worry: You can have very accurate models that work off aggregated data. And, remember, because you are running models for the future, the aggregated models may actually be more accurate than models with a lot of detail that give a false sense of precision.

When you are aggregating, the most important decisions will be around your customers and products.

And if you are going to model in something other than annual buckets, make sure you carefully understand these ramifications.

End-of-Chapter Questions

1. A large electronics and home-appliance retailer is looking to perform a distribution network analysis to optimize its warehouse locations. Using any real electronics retailer that you know as a reference, what are the aggregated product groups that you would consider modeling? Make sure you consider all the factors, including product sourcing, product logistics, characteristics, and so on.

2. Open the file `Aggregating Customers.xls` found on the book Web site. This file contains a list of 10,000 raw ship-to points and some characteristics about these data points. How would you go about aggregating the customers? Go ahead and follow your strategy. How many customers do you end up with?

3. Open the file `Aggregating Products.xls` found on the book Web site. This file contains a list of 50 products. How would you aggregate these products? How few product groups do you need in order to capture all necessary characteristics of these 50 products? Which of the 50 products belong to each of your product groups?

4. You are working for a company with 500 unique products. You must aggregate these products. The team has decided that they will build a simple way to group products by coming up with six categories. Each category only has three choices. They will apply the categories sequentially. For example, they first break the products into Large, Medium, and Small; then for each size they break into Red, Yellow, and Blue; and for each color they break into three more subcategories; and so on for the six categories. What is wrong with this approach?

5. You work for an online retailer and you shipped to 100,000 unique addresses last year. You are building a network model to determine where to open a new warehouse later in the year. Besides the fact that you don't want to build a model with 100,000 ship-to points, why will these 100,000 points not be the same exact points you ship to next year? How different are the points likely to be?

6. You are building a model to determine the best location for a single pharmaceutical plant to make a new drug to serve global demand. Why is modeling every country as a single demand point probably good enough for this problem?

7. You are building a model for a global chemical company that has three plants, each making a unique set of products. They have a plant in Asia, one in North America, and one in Europe. They make hundreds of different products, but the products all have essentially the same size, weight, and handling costs, and ship the same way. What is the minimum number of aggregate products you need? What happens if you have fewer than this number of aggregate products?

14

CREATING A GROUP AND RUNNING A PROJECT

In this chapter, we will discuss how you actually complete network studies within a company. We will cover how you build a group and how you run a project.

Typical Steps to Complete a Network Design Study

In a typical project, you are likely to run into problems with the data as well as organizational challenges of working on a project that impacts many people within a firm. To get a network design study done, you need to treat it as a project and manage it as you would manage any complex project within a company. Of course, there are elements unique to a network design study. In this section, we will cover the typical steps that you want to include in your network design project plan. Broadly, any network design project can be broken into five main steps or phases:

1. Model scoping and data collection phase

2. Data analysis and validation phase

3. Baseline development and validation phase

4. What-if scenario analysis

5. Final conclusion and development of recommendations

Each step is critical and has its own specific purpose. It is important for the project team to go through all the phases, irrespective of the scope and complexity of the supply chain being analyzed or the amount of time available to complete the analysis.

Step 1: Model Scoping and Data Collection

Before you start any project, it is important to first understand the questions that are to be answered and the associated parts of the supply chain that may be impacted. This step may seem trivial and is often overlooked, but it is very important to have a clear understanding of what decisions are being made, and which parts of the supply chain are open to change and which parts are not. In this phase of the project, you are applying the lessons learned from Chapter 12, "The Art of Modeling," and specifically the section "Understanding the Supply Chain."

For a retailer that recently acquired another retail company, the key questions are likely to be:

- What is the optimal combined distribution network that minimizes logistics costs and maximizes service to stores and customers?

- Which existing distribution center locations are redundant and can be closed?

- What is the best way to distribute products to the newly combined store network?

For a consumer-products company that is looking to develop their long-term manufacturing strategy to support growth, the questions would be similar to the following:

- Should we expand existing plants or build new plants? If so, where and when?

- Which products should we manufacture internally and for which products should we use contract manufacturers?

- Is there an opportunity to source products across various regions?

These are just two distinct examples; every supply chain network design study will have a different scope.

What *is* the same between all projects however is that it is critical that you get everyone on the team to agree to the scope and questions the optimization will answer from the start. If you have to go back and make changes to this later, you will likely lose a significant amount of time and may have to effectively start over.

After the scope and questions to answer are finalized, the project team will need to come up with a list of all data that needs to be collected to build the model. This step needs to include a detailed discussion on the level of aggregation and what systems or third party sources the data will come from.

Depending on the model, you will then want to use the knowledge you gained from previous chapters in this book to determine what data goes into this specific model and how you should think about aggregating this data. After you have gone through the exercise of building your list of required data, you now need to collect the data. Collecting the data can often be a very time-consuming and frustrating experience.

Based on our experience, here are some tips to make your data collection efforts more successful and to help you determine how much effort you should anticipate. Note that in this phase we are looking to just access and extract the data. In the next phase we will do a thorough validation of the all the data.

1. Be prepared to collect data from outside a firm's internal systems. There is data that exists in existing systems and data that does not. The purpose of a network design study is to understand the impact of running your supply chain in a different way. This may mean using different plants and warehouses, using new transportation lanes, making products in new locations, and so on. The key is that you need to be prepared to collect data or extrapolate from internal data for new elements you will want to consider in the analysis.

2. If multiple IT systems store the same type of data (say demand), you will need to spend more time collecting and validating the data. There is a chance that the systems will have different fields, that they will have different data definitions, and that the ID fields may not match up.

3. IT systems may be set up for accurate accounting and financial reporting, not necessarily for good supply chain analysis. So your existing systems may not have all the fields you need, and some data might not match up as you would expect.

4. When you gather cost information for new data elements, make sure that it matches with existing data. For example, firms will often have good transportation rates on lanes they use. These rates are often the result of negotiation with the carriers. If you ask for new transportation rates for new lanes, make sure that they are not the "retail" rates or they will be much higher than existing rates. As previously discussed in Chapter 6, "Adding Outbound Transportation to the Model," if retail rates do need to be used, ensure these rates are used for both new *and* old lanes alike.

5. Make sure you understand the accounting cost data before using it. Often, accounting systems will allocate fixed and variable costs to a product in ways that do not make sense for network design studies. See Chapter 7, "Introducing Facility Fixed and Variable Costs," for a thorough discussion of this topic.

6. It is usually better to collect raw data and not aggregated data. Although you have a plan for how you will aggregate, it will not save you time to pull aggregated data from your systems. Most likely, you will want to tweak the aggregation strategies or will need to validate the data that goes into aggregated items. It is best to ask for the raw data and do the simple step of aggregation yourself.

If the data collection is proving to be an extremely difficult task, don't forget to review the lessons learned in Chapter 12 and Chapter 13, "Data Aggregation in Network Design." You may be trying to collect more data than you actually need in order to get the answers to your defined network design questions.

Step 2: Data Analysis and Validation

After the data is collected, it is important to analyze and understand the data to ensure that it is clean and accurately reflects the way the business operates. Because you will be communicating the results of the study to other people in the organization, this phase of the project serves an important purpose of ensuring that you have a good set of data that you can explain to others and that others will agree to.

This phase includes a combination of the following activities: data cleansing, data analysis, data validation, and data aggregation.

DATA CLEANSING

After the initial data is collected, the first step is to review and fix obvious issues. Examples include:

- **Missing or invalid ZIP Codes**—For example, ZIP Codes in New England start with 0 (zero). Excel tends to drop the leading 0 which then either converts the ZIP Code 06457 (Middletown, Connecticut) into 6457, an invalid ZIP Code or, worse, leads someone to think that the location is in Missouri (ZIPs Codes starting with 645 are in Missouri).

- **Shipment data with missing or invalid weights or cube (cube is a transportation term for the cubic size of a shipment)**—For example, there may be truckload shipments showing shipment weights greater than the legal limits (for example, over 45,000 pounds in the U.S.).

- **Order or shipment data with invalid origins or destinations**—For example, the shipment origin may reflect the location of the supplier's headquarters as opposed to the actual plant or warehouse the shipment originated from.

DATA ANALYSIS

After initial data cleansing is completed, the next step is to analyze the data to understand what it is saying. This can be accomplished by creating tables and summaries that aggregate and present data so that it can be evaluated. Summary reports may include:

- Total outbound volume shipped by wareshouse
- Total inbound volume received into each warehouse

- Number of products with active demand and Pareto analysis showing volume breakdown across products

- Total weight shipped by mode and by warehouse

- Total demand by state or region

- Total volume shipped by vendor or plant

- Cost per pound or or some other standard unit of weight (like hundredweight, ton, kilo, and so on) by mode for inbound and outbound shipments

- Average shipment weight by mode for inbound and outbound shipments

These reports will help paint a picture of how the supply chain operates and how products flow. They will also often point out obvious issues with data quality. For example, the total outbound volume shipped by a warehouse should be generally similar to the total inbound volume received by that warehouse in the same timeframe. If there is a large difference, it could be attributed to one of the following:

- Missing inbound or outbound data

- Dramatic inventory buildup or drawdown at specific locations or for specific products

- Incorrect data fields pulled from systems

DATA VALIDATION

After data summaries have been created, the next step is to validate the information with the business stakeholders overseeing the appropriate functions. For example, information summarizing the logistics costs, transportation mode assignments, or average shipment weight should be validated with the Logistics team. Information related to production costs or capital should be validated with the Finance team. This will serve two purposes for the project:

1. It ensures that the data has been validated by the appropriate owners of this information.

2. It gets people from all parts of the organization engaged upfront in the project so that when the results of the analysis are presented, they are more likely to be comfortable with the results and the recommendations because the underlying data was approved by them.

The data validation process may also help identify other issues with the data that may not have been obvious during the original data cleansing, necessitating a revised pull of data. The steps around data cleansing, analysis, and validation need to be repeated until the respective stakeholders feel comfortable that the data portrays a valid representation of their supply chain.

A side benefit of the data validation process is that it may quickly identify areas for improvement in the current supply chain even before any optimization is actually run. For example, a simple table summarizing the total volume and costs for out-of-region shipments from each warehouse can quickly provoke management to assign someone to fix the problems. This may seem surprising but the data validation exercise provides the ability to summarize and visualize existing data that may not be reported on by the firm on a regular basis.

Data Aggregation

After the data has been validated by the appropriate stakeholders, the next step is to start aggregating the data for the purposes of the network design model. After the previous steps are complete, this step should be relatively simple.

Step 3: Baseline Development and Validation

After the data has been validated and aggregated, it is now time to start building the actual model. When it comes to building a model, it is always best to start with a small, simple working model and add complexity incrementally while ensuring feasibility at each step.

The first model is built to represent the historical or as-is state representing how the supply chain operated historically. This model is referred to as the "baseline" model and it serves a couple of purposes:

- It helps validate that the model designed and developed is accurate. Because we are creating a representation of the current supply chain in the software, it is important to be able to compare model outputs against historical financial results for the same input data.

- It serves as the basis for creating additional what-if scenarios in the model.

To prepare for the baseline model build, you need to create the required data tables for importing into the network modeling software. The actual tables and the sequence of data imports may vary depending on the specific software application used.

When building the baseline model, do not think you will load every data element and run. As mentioned previously, it is best to start simple and then add more and more complex data elements as you go. Surprisingly, because this is what makes a baseline a baseline, we have often found that the last thing you want to add to the model is the historical flows of products. After the model is working and all the costs are loaded, you then finally add the historical flows.

It is important to be aware that adding the historical flows can get complicated if your data does not match. For example, let's say that you ship 25.4 million pounds out of your Riverside warehouse and the Bridgewater warehouse ships 45.1 million pounds (see Figure 14.1). Assume that you have validated these shipments and these are the correct numbers you would like to use. However, when you analyze the shipments into the Riverside and Bridgewater warehouses from the Waco and Des Moines plants, you notice that the inbound and outbound numbers do not match up (also shown in Figure 14.1). This may be caused by various things, including poor data quality, accounting procedures, or warehouse transfers. One simple solution is to leave the results as they are. Then, a subsequent version of the baseline would have you make an adjustment to the inbound flows so that they match up with the outbound flows. This is shown in Figure 14.1 where we calculate the actual percentage mix coming from each of the plants and then multiply that by the total outbound shipments from the warehouses to come up with an inbound flow that equals the outbound flow. Remember, when the optimization scenarios run, the results will show an inbound flow that matches the outbound flow. Therefore it is important for us to maintain the same equality in the baseline so we can compare back to it.

Outbound Volume Summary

Origin Whs	Cust Type	Total Wt (MM lbs)
Riverside, CA	All customers	25.4
Bridgewater, NJ	All customers	45.1
TOTAL		**70.5**

Inbound Volume Summary and Adjustment

Origin Plant	Destination Whs	Actual Vol (MM lbs)	% Mix	Adj Inbound Vol (MM lbs)
Des Moines, IA	Riverside, CA	10.1	43%	11.0
Waco, TX	Riverside, CA	13.2	57%	14.4
	Total	**23.3**		**25.4**
Des Moines, IA	Bridgewater, NJ	34.5	70%	31.6
Waco, TX	Bridgewater, NJ	14.8	30%	13.5
	Total	**49.3**		**45.1**

Figure 14.1   Example Showing Balancing of Inbound and Outbound Flows in Baseline Modeling

After the baseline model starts providing reasonable results, the next step is to start comparing the outputs against actual financial results or original systems data from the same time period. The focus is primarily on model cost versus actual costs, but you may also want to validate volumes and capacity utilization. Besides validating the total costs, you may also want to validate the costs of various categories such as:

- Inbound transportation costs by mode
- Outbound transportation costs by mode
- Warehouse fixed costs
- Warehouse variable costs
- Manufacturing variable costs (if applicable)
- Manufacturing fixed costs (if applicable)
- Sourcing costs

The results need to be validated, usually within 1% to 10% of actual costs. After validation, you want to run the appropriate optimized baseline model. For a more detailed discussion on validation and the optimized baseline model, see Chapter 8, "Baselines and Optimal Baselines."

After the baseline and optimized baseline scenarios are completed, it is time to document all the data assumptions, the business rules, and a summary of the baseline model results and present them to the project team and stakeholders. This is a major milestone in the project life cycle and it requires validation and sign-off from all the appropriate sponsors and stakeholders.

Step 4: What-If Scenario Analysis

After the baseline model is developed and approved, we are now ready to start running what-if scenarios. This phase represents the "fun" part of a project, and the phase where the real focus lies in ensuring all the key project questions get addressed.

Before we start running any scenarios, it is important to review the key questions that were stated during the project scoping phase and develop a list of scenarios that would makes sense to run in order to address those questions.

The most important scenarios to run are those that answer the key questions. For example, if the key question is to find the best three, four, and five warehouses, you want to run those scenarios. But, you want to hit the run button more than just three times. To best answer the key questions, you have to understand different what-if questions. A project will have many different what-if questions. Here are three basic ones.

- What if we picked a different set or different number of facilities?
- What if demand was higher, what if demand was lower?
- What if our projected costs were higher, what if they were lower?

The goal with the what-if scenarios is to make sure you have a solid answer, to make sure you can explain your answer, and to make sure the answer is robust.

The more scenarios you run, the better the answer you will have. You will trust that the model has been set up correctly and well-tested, you will have explored many solutions and have a good understanding of the best solution, and you will be able to understand when different solutions have similar costs. It is important to find different solutions with similar costs. There will be many non-quantifiable factors that the team will want to consider. By having a range of solutions to choose from, you can better factor the non-quantifiable costs into the decision-making process.

The more scenarios you run, the better you will be able to explain your answer. The scenario runs allow you to understand what factors are most important to the solutions. For example, what is driving the answer? Is it transportation costs or manufacturing costs or the need to be located close to customers? Remember, we need to explain the results to a wide range of people who are not as familiar with the model or network modeling, in general. The better you can understand and explain the results, the higher the chance that others will understand it as well.

The more scenarios you run, the better you can understand how robust the solutions are. Many of the inputs to a network design model are forecasts or projections. For example, demand and transportation costs next year are not known. You want to run the model to test different forecasts and projections. You want to understand how the answer changes as key input values change. But, also, what you are doing is trying to understand how well your solutions hold up if the forecasts and projections are wrong. For example, a solution that is "great" for one set of input data but terrible if that input changes may not be as good as a solution that is just "good" for one set of input data but still "good" if that input changes.

Also, don't be afraid to run a lot of different scenarios that may not directly address the key questions. You have now seen examples of many different types of scenario runs throughout this book. Feel free to experiment and use your creativity.

We will learn something new with each scenario, which will in turn raise more questions requiring the running of additional scenarios with more specific variations and changes. This is typical in a network design project; it is a healthy process and illustrates the power of the iterative what-if scenario analysis. Ultimately, the model will help us learn what factors are really driving the recommended structure of the network.

Step 5: Final Conclusion and Development of Recommendations

No matter how fun the scenario analysis is, we eventually have to come to a conclusion. So after we have run a sufficient number of scenarios to test various alternatives, and understand the best solutions, it is time to compile the results along with supporting analysis for presenting to the management team.

At a high level, this may seem straightforward given that the hard work of collecting the data, building the model, and running the scenarios is complete. However, this step is often the most important and critical part of the entire project. This is because this phase is where we present and sell the results of the entire analysis—even if the study was based on sound analysis and extensive due diligence, it may end up as a futile exercise if it is not presented with all of the compiled information and recommendations summarized in a concise manner that helps management understand and make better decisions.

When you are developing the final recommendation(s), it is important to consider both qualitative and quantitative factors that are not covered as part of the model as well. These may include:

- Complexity of implementation (related to how many new sites are opened and how many are closed; the higher the number of changes, the higher the complexity)

- Availability of labor and space (if new plant or warehouse sites are recommended)

- Impact of network changes on customer perception and demand

- Timeline and road map for implementation of changes

- Dependence on other factors such as IT system changes/availability, integration with third-party vendors

- Tax and regulatory implications

Following is a high-level overview of a good structure for a final presentation:

- Project Objectives and Scope Review
    - The objectives of the study should be clearly outlined, specifying the questions that are being answered by the analysis.
    - It is equally important to highlight the key questions that are not part of the scope of the study, especially if this study focuses on a subset of the overall initiative. This can help the presentation go much more smoothly, making it easier for the audience to understand the context and to set their expectations appropriately from the start.

- Executive Summary (Optional)
    - When the audience includes senior executives, it is useful to include a one-slide summary of the key findings and recommendations to make the best use of the limited time you are given with them.

- Project and Data Assumptions

  - This section covers a quick summary of the detailed scope and a review of the data assumptions, calling out those that were used to bridge gaps in data.

- Baseline Validation (Optional)

  - It may be beneficial to quickly touch on the baseline model results to remind the audience about this intermediate milestone.

  - It is also important to touch on how the baseline scenario was validated against the financials, and therefore is a legitimate basis to compare against future scenarios.

- Scenario Analysis

  - The first section focuses on the scenarios tied to the main questions that were evaluated as part of the analysis.

  - The second (shorter) section focuses on sensitivity analysis run on the key finalized recommendation(s) to test impact of key variables. It would also make sense to address other qualitative factors such as labor or space availability, implementation complexity, and so on.

- Final Summary and Recommendations

  - The final recommendations section should list the best solution(s) for further evaluation and analysis. Note that this type of exercise is meant to provide decision support to management, and is not intended to yield a single best solution. Depending on the recommended solution(s), additional follow-on validation exercises may be required, including developing a detailed financial business case (with ROI and cash flow analyses), especially if the final solution requires extensive capital investment.

- Next Steps

  - A typical network design study rarely ends with the final presentation—there are often additional scenarios or follow-on analyses as a result of discussions in these meetings.

  - Once these additional scenarios and follow-on analyses are run, and discussed, and decided upon, the project team then develops the implementation plan to turn the strategy into a reality.

Setting Up a Modeling Group

The preceding steps are the ones you will follow for any network design project. An important question we have not yet discussed often arises directly after the initial need for the study has been determined. Should a firm do the project themselves or have a third-party consulting firm perform it on their behalf?

While there is no correct answer to this question, the following guidelines are helpful for making the decision.

It may be better for a firm to do this work internally if any of the following situations is true:

- The firm's supply chain is large, dynamic, and changes frequently. It may be changing through acquisitions or through the need to analyze each of your distinct markets, like North America, South America, Europe, Asia, and so on.

- The firm may have many different divisions or business units that require separate analysis.

- A single firm may have many different network optimization needs such as the ability to do budgeting and capital planning, reconfiguring warehouse territories, planning for seasonal spikes or seasonal changes to their supply chain.

- A firm needs to rerun scenarios or tweak the solutions during the implementation so they can adjust as time passes and the new supply chain takes shape.

Of course, the opposite of the preceding list gives you reasons for using a consulting firm. In addition, there are a few unique items to add to the reasons for working with a consultant:

- Unannounced acquisitions, potential major layoffs, or new product introductions may require dealing with very sensitive data and therefore using third parties to analyze may provide the needed anonymity.

- The firm has an especially hard project and needs to bring in people who are experts at network design models and who can add value in other aspects of the project management as well.

- The firm has a very tight deadline on the timing of the project's completion and therefore needs additional experienced team members quickly to ensure its completion in time. We have seen this occur when companies want to validate a new location before starting to implement a major decision previously made without the validation of network modeling.

- In some cases, the political environment is such that it is best to have a third party work on the project. Unannounced acquisitions, potential major layoffs, or new product introductions may deal with very sensitive data and therefore utilizing third parties to analyze may ensure the needed anonymity.

If a firm decides to build a modeling group in-house, they need to consider the best way to structure and staff the team. The firm should determine whether to structure this group centrally or deploy to the regions and business units. The benefit of a central group is better capturing and sharing of knowledge across the business units and may enable a better balance of the workload generated by many different projects. The benefit of regional groups or groups by business units is that firms can push the knowledge deeper in the organization. Often, large firms do not have to choose between the two extremes though. There can be a small central group that supports the regions and business units. The central group may be called in for the more difficult models and facilitate the sharing of information across the organization.

A firm will also need to determine how they will pay for the group and how it will generate projects. They may want to set up the group as a function that is paid by the corporate office conducting projects that help the entire business. They may also set up the group so that each region or business unit pays for their own support. For projects, there may be mandatory projects that groups need to run, or each business unit may need to come to the group with their own request for an analysis. Also, the group may be responsible for generating demand for their work. However, we have seen many cases in which the initial successes of a group like this creates its own demand through word of mouth within the firm.

A firm then needs to determine how they will staff this group. It is clear that people are needed to do the modeling. The number of modelers however, obviously depends on the amount of work you expect this team to complete. In addition to modelers, the team will also need a manager for the group and/or several project managers to oversee and guide the end to end work of each analysis. The number here again varies with the size of the group, and a person may sometimes play the role of a modeler and a manager at the same time. The modeler, often called an analyst, is responsible for collecting data, validating the data, building the baseline, running scenarios, and assisting with the final presentation. The manager is responsible for running projects, determining the scope, presenting recommendations, and helping the modeler get work done when required. Besides this core group, the team also needs to decide whether full or part time IT experts (for data collection) and subject matter experts (transportation or manufacturing) will be needed on the team to assist in their area of expertise. If these people are not on the team, their help will still be needed for key questions regarding extraction and clarification of modeling data. The larger the group, the more likely the team will have full-time demand for this expertise. Otherwise, it is probably best to share this expertise with other groups.

When thinking about the optimal skillset for people on the modeling team, a firm will want people who are relatively technical and able to work with large data sets. They need to be able to communicate with a wide range of people in the organization and have a good understanding of or willingness to learn the functions across the entire supply chain. Also, equally important, these people must be comfortable with some level of ambiguity. Network modeling is not accounting. There will always be a margin of error in the results and decisions need to be made with confidence in the data and assumptions used in the model. A good deal of time can be wasted when modelers become too intent on very granular levels of data and poor aggregation strategies which inevitably leads to a disastrous project timeline.

Lessons Learned

It is important to follow the methodology for running a network design project in order to ensure that the questions are appropriately answered and the project is successful. These are large projects and touch many people in the organization. Firms need to manage this like a project and not treat this just as a technical exercise.

There are valid reasons for doing this work internally and for having a third party consulting firm do it. If you build the group yourself, it is important that you structure the group in a way that ensures a high likelihood of success from the start.

End-of-Chapter Questions

1. If you are considering closing a warehouse and opening a new one, what people in the organization might be impacted and how might they react to the project?

2. Name other reasons a firm may want to build a group and other reasons they may use a consulting firm. Describe a hybrid approach (there is an in-house capability and a consulting firm is also used to help with the analysis) and the pros and cons of a hybrid approach.

3. When collecting data, you find that the demand data is listed by total units sold to each ship-to location, but the transportation that shows which customer was served by which warehouse only provides the total weight moving between the sites. You would like to be able to match up the files but cannot because you don't have product information in the transportation file. Setting aside your needs, why might this data be perfectly fine for the rest of the organization?

4. When validating the demand information, you discover that the customer information is given as the bill-to address. Why won't this help you?

5. During the scenario analysis stage, why is it important to determine specific scenarios you want to run? Why is it also important to experiment?

6. In the final presentation step, why is it important to review the data collection?

15

CASE STUDY:
JPMS CHEMICALS CASE STUDY

In this chapter, we will look at modeling and analyzing capacity from a manufacturing perspective, as well as how manufacturing and sourcing optimization can be modeled. We will try various what-if scenarios to explore different network alternatives with an example of a chemical company in India.

Indian Chemical Company—Case Study

In this case study, we will review the case of JPMS Chemicals Pvt. Ltd., a large manufacturer of various chemicals. JPMS Chemicals is a privately owned business based in Mumbai, with sales and distribution to other chemical manufacturers and distributors all over India. The company operates a network of four plants located in Madurai (Tamil Nadu), Aurangabad (Maharashtra), Dewas (Madhya Pradesh), and Lucknow (Uttar Pradesh).

The company manufactures and distributes several types of products, including resins and solvents. For the purposes of this modeling exercise, we will simplify them into three product families: Chemical A, Chemical B, and Chemical C. Note that these are the aggregated products that we are modeling—each family includes hundreds of SKUs with similar characteristics.

The map in Figure 15.1 shows the current four plants overlaid against their customer demand points that they ship to. Their customer base includes small distributors as well as manufacturing plants or warehouses, and this aligns with overall population demographics.

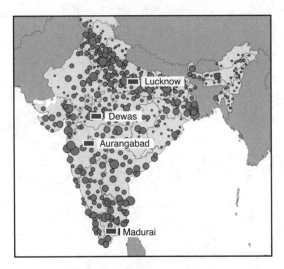

Figure 15.1　Map of Customers and Plant Locations

All three products can be manufactured at all the plants, so they serve as sources of regional supply—that is, each plant produces products to serve its regional customer base. The table in Figure 15.2 shows the maximum production capacity (in units) available by plant by product family. For Chemical A, the maximum output for one production line is 550,000 units, and both Aurangabad and Lucknow plants have one line each for Chemical A. The other two plants have two lines for Chemical A. For Chemical B, all plants except Aurangabad have one line only. For Chemical C, all plants except Dewas have two lines each (capacity of 11,000 units/line), and the Dewas plant has three lines dedicated to Chemical C.

Plant	Number of Lines			Max Capacity by Line		
	Chemical A	Chemical B	Chemical C	Chemical A	Chemical B	Chemical C
Aurangabad	1 line	2 lines	2 lines	550.000	2,200,000	22,000
Dewas	2 lines	1 line	3 lines	1,100,000	1,100,000	33,000
Lucknow	1 line	1 line	2 lines	550.000	1,100,000	22,000
Madurai	2 lines	1 line	2 lines	1,100,000	1,100,000	22,000

Figure 15.2　Production Capacity by Product by Plant

With the ongoing economic boom in India, the management team was projecting aggressive growth in demand over the next few years and was concerned whether there was sufficient manufacturing capacity to support growth. They wanted to understand whether they needed a new plant, and if so where and how big it should be.

When we analyze manufacturing or sourcing strategies, it is important to factor the sourcing of raw materials and components into these plants—that is, we want to consider the impact of inbound transportation and sourcing costs. In this case, the components used to manufacture these chemicals are widely available and can be locally procured even at any new locations at roughly the same cost. Based on this, we are not including raw materials and components in this case study.

Let's start by building a baseline model of the current network with the last 12 months of demand. It always makes sense to start with a baseline model to ensure that the model is accurate and is a good representation of how the supply chain operated.

From Figure 15.3, we can see that the customers are regionally sourced, with some overlap, from the four plants in the network. This has traditionally been the company's sourcing strategy because this allowed them to reduce transportation costs and provide good customer service.

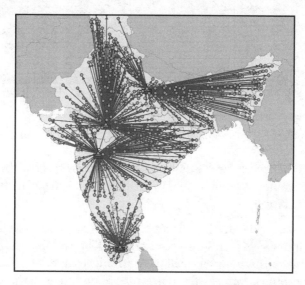

Figure 15.3   Map Showing Results of Baseline Model

The chart in Figure 15.4 shows that three of the four plants are highly utilized without much spare capacity available. There is some capacity available in the Madurai plant, but given this plant is located in the southernmost part of the country, it will be very expensive to ship across the country when other plants are out of capacity. This shows that there will likely be a need for additional capacity to support the aggressive growth plans.

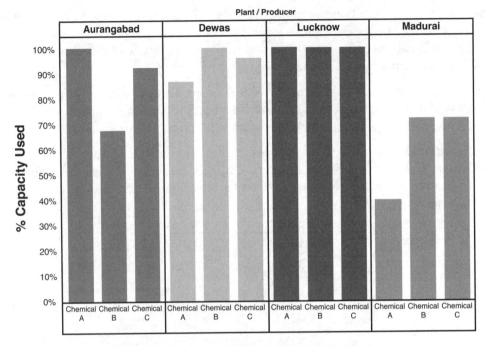

Figure 15.4   Capacity Utilization by Plant by Product

The JPMS management team was estimating the business to grow by 20% within the next 3 years, and wanted to understand what this would mean in terms of available manufacturing capacity.

Let's run a what-if scenario with the current plants and capacity but with a 20% increase in customer demand. We will run this with the overall objective set to maximize profits, rather than minimize costs, in order to allow the model to develop a feasible solution. In a model with maximum profits as the objective, we no longer have to enforce that all demand is met. Because meeting each demand adds to the overall profit (if the revenue is high enough), the model will naturally want to meet all the demand. In this case, we are trying to understand how much demand is unmet. In reality, JPMS wants to meet all the demand. Note: This is a common modeling trick used to troubleshoot a minimum cost model that cannot meet all the demand.

After running the scenario, let's see how the results compared when looking at costs and profits because we ran the growth scenario in profit-maximization mode (see Figure 15.5).

Category	Baseline	Current Network with 20% Demand Growth	% Difference
TOTAL COST (in Million Rupees)	83.2	100.7	21%
Production Cost	70.9	84.9	20%
Outbound Shipping Cost	12.3	15.8	28%
REVENUE	283.4	339.8	20%
PROFIT	200.2	239.1	19%

Figure 15.5 Scenario Comparison—Baseline Demand Versus 20% Demand Growth (Current Network)

The cost figures in Figure 15.5 give us some indication of what the model is doing given the increased demand growth. Firstly, the production cost increased by 20% compared to the baseline. Because we did not change the unit production costs, the increase in production costs can be attributed only to the higher volume (20%) being produced. This also tells us that the current network had sufficient capacity to meet the 20% demand growth. Secondly, we notice that the transportation costs increased by 28% compared to the baseline. We had held the transportation rates constant, so this is not a contributor to the higher freight costs. Also, we had applied the 20% growth uniformly across all customer demand points—so there are no variations in growth demographics that would lead to a disproportionately high increase in transportation costs. This tells us that the larger increase in transportation is due to longer transportation distances compared to the baseline, which could be generated by assigning customer demand to plants where spare capacity is available.

A quick look at the maps for the two scenarios illustrates the capacity situation (see Figure 15.6). In the baseline, we can see that the Madurai plant (South India) serves demand only in South India. This was also the plant with some spare capacity available. When the demand was increased, we see that Madurai is now serving customers in East (Orissa) and Central India.

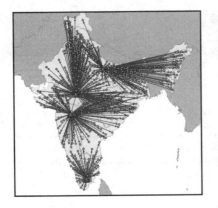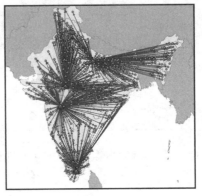

Figure 15.6   Map Comparison—Baseline Demand Versus 20% Demand Growth

When we look at the production capacity utilization by plant and compare the numbers between baseline and the demand growth scenario (see Figure 15.7), this confirms what we see in other reports. We see that the Madurai plant is close to 100% utilized across all three products, whereas Dewas has some spare capacity for Chemical A.

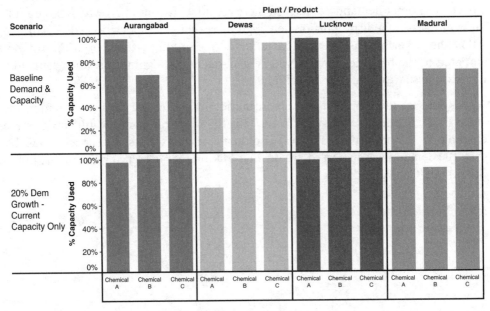

Figure 15.7   Production Capacity Utilization—Baseline Versus Scenario with 20% Growth

When we look at the customer service aspect of this scenario (see Figure 15.8), this shows another key aspect that is important to consider when evaluating new alternatives.

% of Demand Served		
Distance Band (kms)	Baseline	20% Demand Growth
200	12%	11%
800	78%	75%
1200	95%	90%
1600	99%	96%
3500	100%	100%

Figure 15.8 Services Levels (% of Demand Served) by Distance Band

This shows that service levels were similar for the 800km distance bands, but started deteriorating for longer distances. This makes sense as the model would try to serve demand within short distances first, and serve remaining demand from wherever capacity is available.

The preceding scenario tells us and the management team a few things:

- There is capacity available within the current network to support 20% demand growth.

- However, this will mean higher logistics costs in order to utilize available capacity. This would also have a negative impact on customer service levels, especially for customers located beyond 800km from plants.

- This also tells us that there is very limited spare capacity available to handle demand spikes higher than 20%.

Given the capacity sensitivity, the management team wanted to understand the best location(s) to open a new plant. To answer this question, we will add a list of potential plant locations based on customer and demand demographics. For the purpose of this analysis, we will add a list of 30 locations to the model. The list of locations and their placement on a map is shown in Figure 15.9.

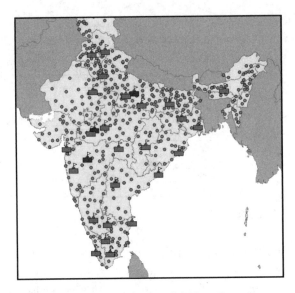

Figure 15.9   List of Potential Plant Locations

For the new plant(s), there are three main questions we are trying to answer:

1. Where should the new plant be located?

2. What products should be manufactured at the new plant(s)?

3. How much capacity needs to be added at the new location(s)—that is, how many production lines do we need to add?

In terms of production capability, we want the model to determine what to make at the new plant(s), so we will give the potential plants the ability to manufacture all products.

As far as capacity is concerned, there are a couple of ways to model this:

1. We could give the new plants unlimited capacity to produce any products. This will allow the model to determine how much product should be produced—we can then determine how many lines are needed based on this.

2. We could add capacity representing a certain number of lines, and let the model utilize this available capacity.

Both options are valid and applicable to model capacity. The best approach will depend on the nature of the business, capital equipment costs, and so on.

The advantage of Option 1 is that it would give the model the most flexibility in determining product sourcing volume from the new plant. Alternatively, this could also be viewed as a disadvantage because the model may come up with a volume mix that represents 1.05 lines—that is, 5% capacity utilized for a second line. It will probably not make sense to open a second production line to handle a very small incremental spillover volume.

Given this, we will use Option 2 and model a specific number of lines at each potential plant and see how the model utilizes this capacity. To give the model sufficient flexibility, we will give each new plant three potential lines for Chemicals A and B, and four potential lines for Chemical C. This allows the new plants to be potentially much larger than the existing plants.

For now, we will not add any opening costs or fixed costs for the new lines—we will look at this a bit later. As we have seen in earlier chapters, we can evaluate the impact of fixed costs without having to model them directly in the model.

Figure 15.10 shows the existing and potential plants with the available lines and capacity.

Plant Type	Supplier	Number of Lines			Max Capacity by Line		
		Chemical A	Chemical B	Chemical C	Chemical A	Chemical B	Chemical C
Existing	Aurangabad	1 line	2 lines	2 lines	550.000	2,200,000	22,000
	Dewas	2 lines	1 line	3 lines	1,100,000	1,100,000	33,000
	Lucknow	1 line	1 line	2 lines	550,000	1,100,000	22,000
	Madurai	2 lines	1 line	2 lines	1,100,000	1,100,000	22,000
Potential	Allahabad	3 lines	3 lines	4 lines	1,650,000	3,300,000	44,000
	Bangalore	3 lines	3 lines	4 lines	1,650,000	3,300,000	44,000
	Bhopal	3 lines	3 lines	4 lines	1,650,000	3,300,000	44,000
	Bhubaneswar	3 lines	3 lines	4 lines	1,650,000	3,300,000	44,000
	Chennai	3 lines	3 lines	4 lines	1,650,000	3,300,000	44,000
	Coimbatore	3 lines	3 lines	4 lines	1,650,000	3,300,000	44,000
	Delhi	3 lines	3 lines	4 lines	1,650,000	3,300,000	44,000
	Dhanbad	3 lines	3 lines	4 lines	1,650,000	3,300,000	44,000
	Guwahati	3 lines	3 lines	4 lines	1,650,000	3,300,000	44,000
	Howrah	3 lines	3 lines	4 lines	1,650,000	3,300,000	44,000
	Hyderabad	3 lines	3 lines	4 lines	1,650,000	3,300,000	44,000
	Indore	3 lines	3 lines	4 lines	1,650,000	3,300,000	44,000
	Jaipur	3 lines	3 lines	4 lines	1,650,000	3,300,000	44,000
	Kanpur	3 lines	3 lines	4 lines	1,650,000	3,300,000	44,000
	Kochi	3 lines	3 lines	4 lines	1,650,000	3,300,000	44,000
	Ludhiana	3 lines	3 lines	4 lines	1,650,000	3,300,000	44,000
	Madurai	3 lines	3 lines	4 lines	1,650,000	3,300,000	44,000
	Mahasamund	3 lines	3 lines	4 lines	1,650,000	3,300,000	44,000
	Mysore	3 lines	3 lines	4 lines	1,650,000	3,300,000	44,000
	Nagpur	3 lines	3 lines	4 lines	1,650,000	3,300,000	44,000
	Patna	3 lines	3 lines	4 lines	1,650,000	3,300,000	44,000
	Pune	3 lines	3 lines	4 lines	1,650,000	3,300,000	44,000
	Shimla	3 lines	3 lines	4 lines	1,650,000	3,300,000	44,000
	Surat	3 lines	3 lines	4 lines	1,650,000	3,300,000	44,000
	Vizianagaram	3 lines	3 lines	4 lines	1,650,000	3,300,000	44,000

Figure 15.10   Production Capacity by Plant for Current and Potential Plants

We will also need to add transportation rates and lanes from the new plants (and attached warehouses) to all the customers. We will also create a plant site grouping for all potential plants, allowing us to apply constraints on the number of new plants that can be opened.

We are now ready to run our new set of scenarios with the new plant locations. The first scenario we will run would be to allow the model to *select up to one new plant* location.

From the map shown in Figure 15.11, we can see that the model has opened a new plant in Dhanbad (Jharkhand) that is serving customers in the eastern part of the country, and the Lucknow plant is serving customers in the northern region only.

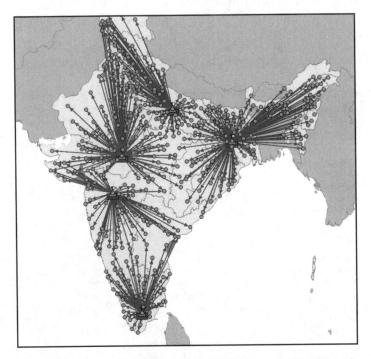

Figure 15.11   Map Showing Results of Scenario with One New Plant

When we look at the costs comparison (see Figure 15.12) against the scenario without new plants, we see that the overall costs reduced by 5%—the savings are attributed primarily in outbound transportation costs because the model now has sufficient capacity to serve demand regionally.

Category	Future Demand - 1 New Plant	Future Demand - No New Plants	% Difference
TOTAL COST (in Million Rupees)	95.9	100.7	**5%**
Production Cost	85.0	84.9	**0%**
Outbound Shipping Cost	10.8	15.8	**31%**
REVENUE	340.1	339.8	**0%**
PROFIT	244.3	239.1	**-2%**

Figure 15.12   Cost Comparison for Scenarios With and Without New Plant

The production capacity analysis (see Figure 15.13) shows that at the new plant the model opened two production lines each for Chemicals A and B, and three lines for Chemical C. We can see that three of the seven lines opened are highly underutilized. This is because the model did not have any penalties to open a new line, so it opened a new line if it was cheaper to serve even only a handful of customers from the new plant. We know that it would not make sense to open a new line that is only 5% to 10% utilized. We could easily address this problem by adding a fixed opening cost for each line.

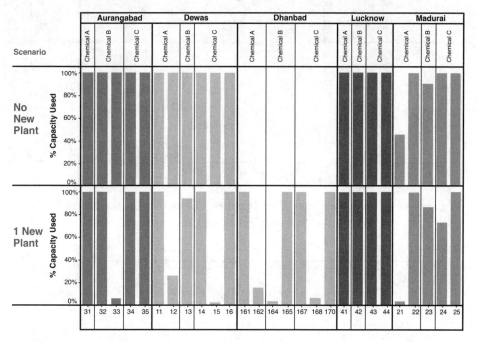

Figure 15.13   Comparison of Production Capacity Utilization Between Scenarios

We will add a fixed cost of Rs 5 million for each line based on the average cost for a new line.

The results of this scenario are shown in Figures 15.14 and 15.15. The chart in Figure 15.14 shows that only one line is opened in the new Dhanbad plant when we add a fixed line cost. The volume served by Dhanbad in the previous scenario is now picked up by Aurangabad and Dewas plants as we see their capacity utilization going up significantly. The consequence of this shift is seen in the cost comparison table. We see that the outbound transportation cost goes up by almost Rs 5 million when we add the fixed line cost at Dhanbad.

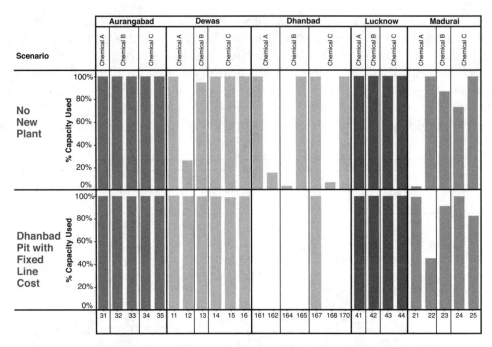

Figure 15.14   Production Capacity Utilization Comparison—New Plant With and
Without Fixed Line Costs

Category	1 New Plant - Fixed Line Costs	1 New Plant	% Difference
TOTAL COST (in Million Rupees)	105.8	95.9	-10%
Production Variable Cost	85.0	85.0	0%
Line Fixed Costs	5.0	0.0	
Outbound Shipping Cost	15.8	10.8	-45%
REVENUE	340.1	340.1	0%
PROFIT	234.3	244.3	4%

Figure 15.15   Scenario Results Comparison—New Plant With and Without Fixed Line Costs

Note that comparing these two scenarios is not an "apples-to-apples" comparison because they do not have the same cost structure. However, we are doing this to help gain insight on the impact of adding a new cost component on the solution as previously developed. The management team can use this insight to understand the key drivers in their supply chain, and thereby make more informed decisions.

Single-Sourcing

The management team had received feedback from some of their customers that they would prefer to receive all their shipments from one plant only. As we can see in the baseline map, there are several customers that are served from more than one plant. This multisource customer assignment was also turning out to be a challenge operationally for the logistics department. So the management team was interested in understanding the impact of single-sourcing individual customer locations.

Let's run a new scenario allowing an additional plant but with the constraint that customers must be single sourced.

The map in Figure 15.16 shows that the new plant in Dhanbad is still valid. The scenario costs (see Figure 15.17) show that there is a less than 1% difference in total costs due to the single-sourcing constraint. This shows that the new network is capable of handling customer single-sourcing with relatively no impact on logistics costs. This has a very positive impact on operational complexity because each customer can be assigned to a plant for order processing and shipping.

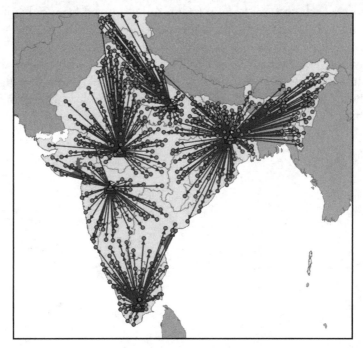

Figure 15.16   Map Showing Results of Scenario with Customers Single-Sourced

Category	1 New Plant - Customer Single-source	1 New Plant	% Difference
TOTAL COST (in Million Rupees)	96.0	95.87	0%
Production Cost	85.0	85.03	0%
Outbound Shipping Cost	11.0	10.84	-1%
REVENUE	340.1	340.1	0%
PROFIT	244.1	244.3	0%

Figure 15.17   Results Comparison of Scenarios With and Without Customer Single-Sourcing

State-Based Single-Sourcing

To evaluate certain tax policies and their implications, the management team wanted to understand the impact of serving each individual state from one plant only—that is, single-sourcing of all customer demand within each state.

To do this, we have created customer groupings by state and applied a constraint in the model that forces each customer group (state) to be sourced from only one plant. The map in Figure 15.18 shows the results of the model with each state single-sourced from one plant/warehouse only. First of all, we notice that the new plant location is not Dhanbad anymore—it has moved east to Howrah instead. Also, we see that the customer assignments look a bit odd—the Madurai plant takes over serving some customers much further North than its previous territory, as well as the new Howrah plant serving customers all the way in the Northwest.

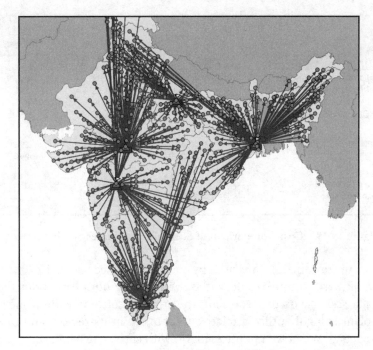

Figure 15.18   Map Showing Results of Scenario with States Single-Sourced

Why is this happening? As discussed in Chapter 5, "Adding Capacity to the Model," we are creating a harder knapsack problem when we aggregate demand into larger chunks. In this case, the model is forced to look for a solution in which each plant supplies the entire demand of the states it serves. Because capacity is tight, we may get answers that look strange.

Even worse than a map that looks strange, we can see from Figure 15.19 that with this constraint JPMS can only meet 98% of the demand. The capacity constraints combined with the constraint to serve a state from one plant do not allow the flexibility to meet all demand. This analysis gives the management team at JPMS some idea of what kind of impact this constraint may have on their supply chain.

Category	Dhanbad Plant with Fixed Costs	1 New Plant with States Single-Sourced	% Difference
TOTAL COST (in Million Rupees)	105.8	104.88	-1%
Production Variable Cost	85.0	83.53	-2%
Line Fixed Costs	5.0	10.00	
Outbound Shipping Cost	15.8	11.35	-39%
REVENUE	340.1	334.1	-2%
PROFIT	234.3	229.3	-2%
Demand Served (Million Units)	8.503	8.353	
% Demand Served	100%	98%	

Figure 15.19 Cost Comparison of Scenario with States Single-Sourced

A quick look at the production capacities by plant by line (see Figure 15.20) shows a picture that is a bit hard to understand without reminding ourselves about the sourcing constraints. The scenario on top represents the previous solution with the new Dhanbad plant. Each plant is highly utilized relative to demand in its region, and one new line opened at Dhanbad. When we ran the scenario with each state being single-sourced, we see that some of the lines are fully utilized whereas others are hardly used. Why is this happening? First of all, the demand distribution across the states is not similar for all products—that is, some states have higher demand for Chemicals A and B and very little demand for C. The model is factoring this constraint, along with transportation costs, and fixed line costs to come up with the capacity utilization profile.

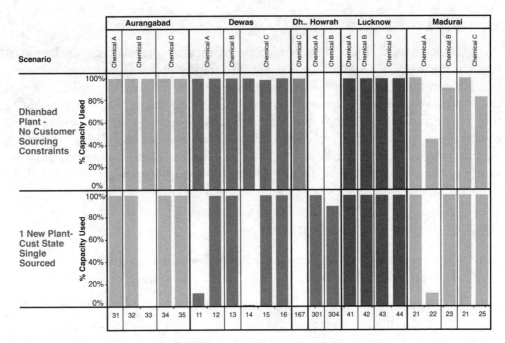

Figure 15.20 Production Capacity Comparison of Scenarios With and
Without Sourcing Constraints

To reiterate our point from earlier, we see that this single-sourcing constraint added a high level of complexity to the model. At the same time, it also provides us with an opportunity to understand the dynamics of the supply chain and how different variables impact each other.

Based on the multiple scenarios run so far, as well as others run to test other potential plant locations, the management team had sufficient information needed to make a sound decision on their manufacturing strategy. Note that we have focused primarily on key supply chain costs that were considered part of the network model—there are several other qualitative factors that impact the choice of a new plant, such as tax benefits, proximity to key customers, availability of skilled labor, and so on.

The management team will need to consider these factors before they make a decision; however, they can combine these non-quantifiable factors with good cost and service data to understand the value of these the non-quantifiable factors.

Lessons Learned from the Case

We developed a model focused on manufacturing capacity and analyzing the best location for a new plant. We could see that capacity constraints were important to understand the key questions that we were trying to answer. However, the capacity constraints can also make the model far more complex, thereby yielding results that may not be intuitive or make sense. This is especially applicable when we start applying sourcing constraints at the aggregate level (for example, by single sourcing a state). In addition, we also reviewed the importance of modeling fixed line costs and how this impacts decisions regarding how many new lines are needed to support demand.

End-of-Chapter Questions

1. If we were to look at two new plant locations, where would they be? (See JPMS.zip on the book Web site for the model and more information.)

2. If we were to look at two new plant locations, where would they be?

3. The management team is interested in evaluating whether it made sense to expand the existing plants versus building a new plant.

    a. How would you go about answering this question?

    b. Can you intuitively identify which plants likely need expansion without running any scenarios? If so, how?

    c. What plants should be expanded and what should be made at those locations?

4. To validate the robustness of the new plant solution, the management team wants to evaluate the 20% growth assumption and its impact on the solution. Run scenarios with −10% (reduction), 10%, 20%, 30%, and 40% growth.

    a. What is the maximum growth that the current network can handle?

    b. How does this impact the selection of the new plant location? Does this vary by product type?

INDEX

C

candy bar supply chain, 3
capacity, fixed costs (facilities), 133
capacity modeling, 83
    constraints, 86-87
    manufacturing capacity, 85-86
    warehouse capacity, 84-85
    weighted average distance problems, 87-89
        constraints, 89-90, 93-96
carbon emissions, geography, 6
categorization of SKUs, 251
categorizing fixed and variable costs by
  analyzing accounting data, 134-135
caves, 10
center of gravity models, 23
center of gravity problems, 23-24
    outbound transportation costs, 101-102,
    118-120
        demand, 102-103
        estimating, 109-111
        multistop costs, estimating, 113-118
        per unit cost, 103-109
        regression analysis, 112-113
    physics weighted-average centering, 24-31
    practical center of gravity, 31, 33-34
central warehouses, 9
co-manufacturers, 11
co-packers, 11
combinations, permutations versus, 44-48
commercial truckload (TL), 104
competitors, non-quantifiable data, 18
components, 196
computational reduction, 91-93
computing weighted-average positions, 29
conservation of flow, 183
consolidating products, warehouses, 7
constraints, 13
    capacity modeling, 86-90, 93-96
    defined, 49
    distance-based facility location problem,
    50-51
    service-level analysis, 75-77
constraints of optimization, fixing, 233
consulting firms, 272
consumer products companies, schematic of
  supply chains, 219
contract manufacturers, 11
converging baseline model to optimization
  model, 141
cost types, aggregation, 258

costs
    per-unit production costs, aggregation, 254
    transportation costs, 99-101
        inbound, 100
        by mode of transportation, 104-107
        outbound, 100-116, 118-120
cross-dock, 8
customer-service level, 208
customers
    aggregation of, 239-242
        validating strategies, national examples,
        242-244
        validating strategies, regional examples,
        244-249
    creating, 145-146
    defined, 38-39
    determining warehouse locations with fixed
    customers, 160-164

D

data, 13
    defined, 48
    demand data, 16
    non-quantifiable data, 17-19
        organizational challenges, 19-21
    precision versus significance, 14-17
    transportation costs, 16
data aggregation, 266
data analysis, 264
    validating, 264-266
data cleansing, 264
data collection, 262-264
data of optimization, fixing, 234
data validation, 265
debugging models, 228-229
decision making, art of modeling, 226-227
decision variables, 13
decisions, 13
    defined, 49
decisions of optimization, fixing, 233
dedicated fleet, 104
demand
    creating, 145-146
    defined, 40
    outbound transportation costs, 102-103
    time horizon, 41-42
    units of measure, 41
demand data, 16
differing service-level requirements, product
  modeling, 179
dimensions of products, impact on transportation, 185-187
disruption costs, 18

outbound flow, distribution network analysis, 187-191

outbound transportation costs, 100-102, 118-120
- demand, 102-103
- estimating, 109-111
- multi-stop costs, estimating, 113-118
- per unit cost, 103-109
- regression analysis, 112-113

P

packaging requirements, aggregation, 253

parcel transport, 105

Pareto optimal solutions, 209-214

per unit cost, outbound transportation, 103-109

per-unit production costs, aggregation, 254

permutations, combinations versus, 44-48

pharmaceutical companies, supply chain schematics, 220

physics weighted-average centering, center of gravity problems, 24-31

plant-attached warehouses, 9

plant capacity modeling. *See* manufacturing capacity modeling

plant locations, source of raw materials, 171

plants, 11
- determining warehouse locations with fixed plants, 160-164
- economies of scale, 10
- production processes, 11
- reasons for having multiple plants, 10
- service levels, 10
- taxes, 11
- transportation costs, 10

practical center of gravity problems, 31-34
- distance-based location. *See* distance-based facility location problems
- outbound transportation costs, 101-102, 118-120
  - *demand, 102-103*
  - *estimating, 109-111*
  - *multi-stop costs, estimating, 113-118*
  - *per unit cost, 103-109*
  - *regression analysis, 112-113*

precision, accuracy versus, 102-103

predefined flows, modeling historical as, 146-147

predefined product families, aggregation, 254

private fleet, 104

product modeling, 177-178
- BOMs, 197
  - *beer manufacturing process modeling, 198-200*
- differing service-level requirements, 179
- mathematical fomulations, 180-184
- product sourcing, 193-195
- Value Grocers, 184-185
  - *distribution network analysis based on outbound flow, 187-191*
  - *impact of product dimensions on transportation, 185-187*
  - *storage restrictions for temperature-controlled products, 191-193*
- variations in logistics characteristics, 178-179

product sourcing, 193-195

production lot sizes, warehouses, 7

production processes, plants, 11

production requirements, aggregation, 253

products
- adding, 201
- aggregation of, 249-250
  - *packaging requirements, 253*
  - *per-unit production costs, 254*
  - *predefined product families, 254*
  - *production requirements, 253*
  - *products that share components or raw materials, 253*
  - *products that share transportation requirements, 254*
  - *removing products with low volumes, 251-252*
  - *size of products, 252-253*
  - *source of products, 250-251*
  - *testing strategies, 254-256*
- dimensions of, impact on transportation, 185-187
- storage restrictions, temperature-controlled products, 191-193

Q

quality, data, 14-17

quantitative data, accuracy, 14-17

R

rail transport, 106

rate matrix, regression analysis, 112-113

raw materials, plant locations, 171

recommendations, development of, 269-270

reduction, 91-93

regional supply, 278

regional warehouses, 9

regression analysis, 112-113

reindustrialization, 208

removing products with low volumes, aggregation, 251-252
results, baseline models, 149-151
retailers, schematics of supply chains, 218
risk
    geography, 6
    nonquantifiable data, 18

S

sales teams, 19
scenarios, running lots of scenarios, 224-225
schematics of supply chains
    for consumer products companies, 219
    for pharmaceutical companies, 220
    for typical retailers, 218
sensitivity analysis, 77-80
service level, geography, 6
service-level analysis, 65-67, 69
    weighted average distance problems, 69-71
      constraints, 75-77
      objective function, 71-74
service-level requirements, product modeling, 179
service levels, 63-64
    measuring, 64-65
    plants, 10
    warehouses, 7
shadow prices, 77
shipment data, analyzing, 145-146
single-sourcing, 289-290
    state-based, 290-293
site aggregation, 256
size of products, aggregation, 252-253
skills, geography, 6
SKUs, categorization of, 251
source of products, aggregation, 250-251
spokes, warehouses, 9
state-based single-sourcing, 290-293
steps to complete a network design study, 261
    step 1: model scoping and data collection, 262-264
    step 2: data analysis and validation, 264-266
    step 3: baseline development and validation, 266-267
    step 4: what-if scenario analysis, 268-269
    step 5: final conclusion and development of recommendations, 269-270
storage restrictions for temperature-controlled products, 191-193
strategic network design, 207
suppliers, 11

supply chain network design, 1
    evaluating, 3-5
    value of, 1-3
supply chains, 218-222
    art of modeling, including things in models that don't exist in the actual supply chain, 225-226
    multi-echelon supply chains, mathematical formulations, 164-166
    schematics
      for consumer products companies, 219
      for pharmaceutical companies, 220
      for typical retailers, 218
    separating the important from the trivial, 222-223
    three-echelon supply chains. *See* three-echelon supply chains

T

tablet supply chain, 3
tax rebates, non-quantifiable data, 18
taxes
    geography, 6
    plants, 11
teams
    finance teams, 20
    logistics teams, 19
    operations teams, 19
    sales teams, 19
temperature-controlled products, storage restrictions, 191-193
testing product aggregation strategies, 254-256
third-party manufacturing sites, 11
three-echelon supply chains, 157
    determining warehouse locations with fixed plants and customers, 160-164
    JADE Paint and Covering, 157-160, 166-171
    linking together, 172-174
time period aggregation, 257-258
TL (truckload), 161
toll manufacturers, 11
total cost versus upfront costs, 208
transportation
    modeling in baselines, 147-148
    product dimensions, 185-187
    products, aggregation, 254
    raw materials, 171
transportation costs, 16, 99-101
    geography, 5
    inbound, 100
    by mode of transportation, 104-107

FINANCIAL TIMES

In an increasingly competitive world, it is quality of thinking that gives an edge—an idea that opens new doors, a technique that solves a problem, or an insight that simply helps make sense of it all.

We work with leading authors in the various arenas of business and finance to bring cutting-edge thinking and best-learning practices to a global market.

It is our goal to create world-class print publications and electronic products that give readers knowledge and understanding that can then be applied, whether studying or at work.

To find out more about our business products, you can visit us at www.ftpress.com.